Introduction to Engineering Design

Synthesis Lectures on Engineering, Science, and Technology

Each book in the series is written by a well known expert in the field. Most titles cover subjects such as professional development, education, and study skills, as well as basic introductory undergraduate material and other topics appropriate for a broader and less technical audience. In addition, the series includes several titles written on very specific topics not covered elsewhere in the Synthesis Digital Library.

Geometric Programming for Design Equation Development and Cost/Profit Optimization (with illustrative case study problems and solutions), Third Edition
Robert C. Creese
December 2016

Engineering Principles in Everyday Life for Non-Engineers
Saeed Benjamin Niku
February 2016

A, B, See... in 3D: A Workbook to Improve 3-D Visualization Skills
Dan G. Dimitriu
August 2015

The Captains of Energy: Systems Dynamics from an Energy Perspective
Vincent C. Prantil and Timothy Decker
February 2015

Lying by Approximation: The Truth about Finite Element Analysis
Vincent C. Prantil, Christopher Papadopoulos, and Paul D. Gessler
August 2013

Simplified Models for Assessing Heat and Mass Transfer in Evaporative Towers
Alessandra De Angelis, Onorio Saro, Giulio Lorenzini, Stefano D'Elia, and Marco Medici
July 2013

The Engineering Design Challenge: A Creative Process
Charles W. Dolan
March 2013

The Making of Green Engineers: Sustainable Development and the Hybrid Imagination
Andrew Jamison
March 2013

Crafting Your Research Future: A Guide to Successful Master's and Ph.D. Degrees in Science & Engineering
Charles X. Ling and Qiang Yang
May 2012

Fundamentals of Engineering Economics and Decision Analysis
David L. Whitman and Ronald E. Terry
April 2012

A Little Book on Teaching: A Beginner's Guide for Educators of Engineering and Applied Science
Steven F. Barrett
March 2012

Engineering Thermodynamics and 21st Century Energy Problems: A Textbook Companion for Student Engagement
Donna Riley
October 2011

MATLAB for Engineering and the Life Sciences
Joseph V. Tranquillo
July 2011

Systems Engineering: Building Successful Systems
Howard Eisner
July 2011

Fin Shape Thermal Optimization Using Bejan's Constructal Theory
Giulio Lorenzini, Simone Moretti, and Alessandra Conti
April 2011

Geometric Programming for Design and Cost Optimization (with illustrative case study problems and solutions), Second Edition
Robert C. Creese
August 2010

Survive and Thrive: A Guide for Untenured Faculty
Wendy C. Crone
August 2010

Geometric Programming for Design and Cost Optimization (with Illustrative Case Study Problems and Solutions)
Robert C. Creese
2009

Style and Ethics of Communication in Science and Engineering
Jay D. Humphrey and Jeffrey W. Holmes
2008

Introduction to Engineering: A Starter's Guide with Hands-On Analog Multimedia Explorations
Lina J. Karam and Naji Mounsef
2008

Introduction to Engineering: A Starter's Guide with Hands-On Digital Multimedia and Robotics Explorations
Lina J. Karam, and Naji Mounsef
2008

CAD/CAM of Sculptured Surfaces on Multi-Axis NC Machine: The DG/K-Based Approach
Stephen P. Radzevich
2008

Tensor Properties of Solids, Part Two: Transport Properties of Solids
Richard F. Tinder
2007

Tensor Properties of Solids, Part One: Equilibrium Tensor Properties of Solids
Richard F. Tinder
2007

Essentials of Applied Mathematics for Scientists and Engineers
Robert G. Watts
2007

Project Management for Engineering Design
Charles Lessard, Joseph Lessard
2007

Relativistic Flight Mechanics and Space Travel
Richard F. Tinder
2006

© Springer Nature Switzerland AG 2022
Reprint of original edition © Morgan & Claypool 2021

Introduction to Engineering Design
Ann Saterbak and Matthew Wettergreen

ISBN: 978-3-031-00965-5 print
ISBN: 978-3-031-02093-3 ebook
ISBN: 978-3-031-00165-9 Hardcover

DOI 10./978-3-031-00165-9

A Publication in the Springer series
SYNTHESIS LECTURES ON ENGINEERING, SCIENCE, AND TECHNOLOGY, #16
Series Editor: Stuart Sabol, Power Engineering

Series ISSN: 2690-0300 Print   2690-0327 Electronic

Introduction to Engineering Design

Ann Saterbak
Duke University
Matthew Wettergreen
Rice Univerity

SYNTHESIS LECTURES ON ON ENGINEERING, SCIENCE, AND TECHNOLOGY #16

ABSTRACT

Introduction to Engineering Design is a practical, straightforward workbook designed to systematize the often messy process of designing solutions to open-ended problems.

From learning about the problem to prototyping a solution, this workbook guides developing engineers and designers through the iterative steps of the engineering design process. Created in a freshman engineering design course over ten years, this workbook has been refined to clearly guide students and teams to success. Together with a series of instructional videos and short project examples, the workbook has space for teams to execute the engineering design process on a challenge of their choice. Designed for university students as well as motivated learners, the workbook supports creative students as they tackle important problems.

Introduction to Engineering Design is designed for educators looking to use project-based engineering design in their classroom.

KEYWORDS

engineering design, decision-making, project-based, teaming, flipped instruction, prototyping, communication

xi

Contents

Acknowledgments xiii

Step 0: Introduction to the Engineering Design Process 1

Step 1: Clarify Team Assignment 9

Step 2: Understand the Problem and Context 21

Step 3: Define Design Criteria 35
    Step 3A: Establishing Objectives and Constraints 36
    Step 3B: User-Defined Scales 43
    Step 3C: Pairwise Comparison Chart 52

Step 4: Brainstorm Solution Options 61
    Step 4A: Decomposition 62
    Step 4B: Brainstorming 68

Step 5: Evaluate Solutions 83
    Step 5A: Pugh Screening 84
    Step 5B: Morph Chart 94
    Step 5C: Pugh Scoring 101

Step 6: Prototype Solution 113
    Step 6A: Safety 114
    Step 6B: Initial Prototyping 120
    Step 6C: Refined Prototyping 132

Step 7: Test Solution 143

Step 8: Finalize a Solution 155

Appendix A: Using this Workbook 165

Appendix B: Teaming 171
    Section B1: Team Orientation 172
    Section B2: Team Pit Stop 179
    Section B3: Team Postmortem 182

Appendix C: Communication 187
    Section C1: Oral Presentation 188
    Section C2: Technical Memos 198

Appendix D: Project Planning 207
    Section D1: Work Breakdown Structure 208
    Section D2: Planning Tools 214

Authors' Biographies 223

Acknowledgments

The authors would like to thank faculty, staff, and students at Rice University and Duke University, specifically Christina Rincon, Alex Nunez-Thompson, Horatia Fang, Zoe Roberts, Dr. Maria Oden, Dr. Marcie O'Malley, and Dr. Sophia Santillan who helped prepare the materials. Others who have helped with this workbook include Dr. Tracy Volz, Adam White, Mike Svat, Amber Muscarello, Dr. Liz Paley, and the members of Houston We Have Coffee, Safe Soap, and IV Drip teams.

Thank you to all of the design teams over the years who have taken first-year design at Rice University and Duke University and provided feedback on the material.

CHAPTER 0

Step 0: Introduction to the Engineering Design Process

Engineers develop comprehensive solutions to difficult problems by following a methodical process of thought, research, and production. Although there are many versions of the engineering design process (EDP), they all contain two main components: the design analysis stage and the solution stage. In the design analysis stage, engineers hold meetings with clients and potential users to develop a conceptual understanding of the problem they are trying to solve. They will then explore the problem through research, interviews, and site visits. In the solution stage, the conceptual understanding is translated into a physical object or system for implementation. It is important to note that the EDP is an *iterative* process, which means several steps repeat to achieve the best possible solution. The EDP sequence is shown in Figure 0.1.

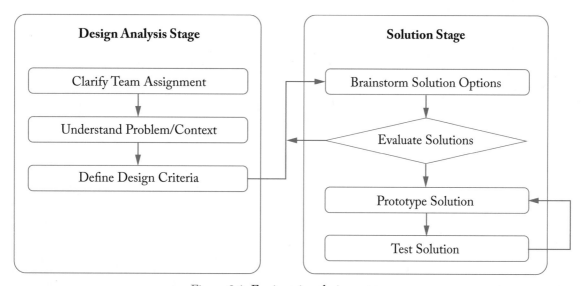

Figure 0.1: Engineering design process.

By the end of this section, you should be equipped to:

- Identify the goals of each step in the EDP,
- Understand the role of the EDP in problem solving, and
- List key characteristics of the EDP.

Playlist

Watch this video playlist for an overview of the EDP. Learn what tasks correspond to each step, as well as characteristics of the overall process.

| Videos | bit.ly/step0-introtoedp |

Video Notes

What are the two stages of the EDP? What is the purpose of each stage?

For each step of the EDP, list an example task in Table 0.1.

Table 0.1: Identifying tasks in the EDP	
Step of EDP	**Example Task**
Clarify Team Assignment	
Understand Problem/Context	
Define Design Criteria	
Brainstorm Solution Options	
Evaluate Solutions	
Prototype Solution	
Test Solution	

Content Summary

The Role of Design in Engineering Practice

Engineering is the application of scientific knowledge to solve problems. Unlike scientific research that uncovers new knowledge, engineering design is the assembly of existing knowledge to produce something that is new. Using the EDP, engineers often produce physical objects, digital systems, or processes.

Engineers can be distinguished from other professionals by their ability to solve complex problems and implement solutions in cost-effective and practical ways. This ability to face a problem, work through various thoughts and abstract ideas, and then translate them into reality is what is so exciting about engineering. There are enough differences between different flavors of engineering that they exist as separate fields: civil, biomedical, chemical, mechanical, electrical, petroleum, environmental, aeronautical, computer, and more.

Engineers solve problems using a number of tools. One important tool is the EDP, which is the focus of this workbook. This process can be applied to small, large, old, or new problems. The process is iterative and decision based, meaning the steps can be applied many times to revise and solve large, world-wide, critically important issues.

Steps of the EDP

Engineering design is an iterative problem-solving process that creates solutions to meet client needs. A model for this process is shown in Figure 0.1. The seven steps of the EDP are as follows:

1. Clarify Team Assignment. Develop a clear understanding of the design challenge by having conversations with the client.

2. Understand Problem/Context. Research the problem space by learning about the existing solutions, relevant background, aspects that govern the problem, and business perspectives. This is achieved by conducting research, talking with users and clients, and traveling to specific destinations.

3. Define Design Criteria. Formally define design goals by pairing quantitative numbers to objectives and constraints.

4. Brainstorm Solution Options. Identify as many solution options as possible using a disciplined process that allows team members to build off of each other's ideas and think divergently.

5. Evaluate Solutions. Reduce a large number of solution options down to a selected design solution.

6. Prototype Solution. Detail the solution and build a CAD model, physical prototype, or computer program. Iterate physical prototypes from low-cost, low-fidelity materials to high-fidelity materials that resemble the solution.

7. Test Solution. Test to see how the design solution meets the established design criteria.

Review Questions

1. Engineering design is a(n) _____ process.

   A. Streamlined

   B. Iterative

   C. Archaic

   D. Linear

2. The two main stages of the EDP are first: _____ and second: _____.

   A. Design analysis stage; solution stage

   B. Solution stage; design analysis stage

   C. Brainstorm solution options stage; build prototype stage

   D. Research problem stage; research solution stage

3. During the design analysis stage teams _____.

   A. Immerse themselves in analyzing solution options

   B. Immerse themselves in understanding the problem

   C. Often rush

   D. Both B and C

4. Knowledge of the EDP allows you to _____.

   A. Take on challenging projects

   B. Work on a variety of projects in a variety of settings

   C. Make informed decisions throughout your design project

   D. All of the above

5. Identify which step should replace the # 2 in the EDP diagram below.

Engineering Design Process

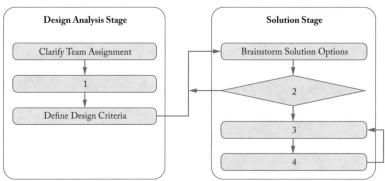

   A. Evaluate solutions

   B. Test solution

   C. Understand problem/context

   D. Prototype solution

Exercise #1

Introduction

Individuals who work in large warehouses are responsible for scanning thousands of barcodes each day. Using standard, trigger-activated scanners can cause stenosing tenosynovitis or "trigger finger," which can be painful and require surgery to remove.

A new engineering team is tasked with designing a new scanner to prevent trigger finger entirely. The team writes out everything they need to accomplish.

Task

Using your knowledge of the EDP, organize the engineering team's notes into the steps of the EDP.

Develop a functioning prototype of the new UPC scanner.

Conduct research about UPC scanners and trigger finger, including interviewing potential users.

Test our device with potential future users for feedback.

Generate many possible solutions for what the new UPC scanner could look like and how it could work.

Set quantitative goals and constraints about what the new UPC scanner must achieve.

Interview the client to better undertand the scope of the project and their ideal outcomes.

Select the best possible design for the new UPC scanner based on all generated ideas.

Exercise #2

Introduction

Your team has been tasked with designing a new feeding device for children with musculoskeletal disorders. Some patients with such disorders have a limited range of motion in their arms and hand, as well as limited gripping power. More information on this project is given in Step 1: Clarify Team Assignment.

Task

Using your knowledge of the EDP, read the following scenarios and identify steps that the team skipped or executed incorrectly. If a scenario begins in the middle of the EDP, assume that the previous steps were completed correctly.

1. Your team reads the project prompt carefully and takes notes. Everyone then splits up the information and begins to research the problem. Once the research has been documented and shared, the team sets design criteria for their final design.

2. After researching the design problem, the team begins to brainstorm solutions. Elements of solutions that were identified while researching were incorporated into the brainstorming process.

3. Once the team has chosen a final design solution, they build prototypes. The first prototype contains parts made from a laser cutter and 3D printer. This initial prototype is then tested and evaluated against the team's design criteria.

4. After brainstorming a wide selection of solutions, the team unanimously votes on a solution. They inform the client of their selection and begin to prototype the solution.

Applying This Step to Your Project

Using this workbook assumes that you are working alone or in a team to solve an open-ended design challenge. This design challenge should be defined to some degree, either in a written document or a conversation. You should know who the client for the problem is, what the solution needs to do, and some objectives for the solution to achieve. Learning the EDP is more fun and efficacious when you have your own problem to tackle.

CHAPTER 1

Step 1: Clarify Team Assignment

Clarifying the team assignment is the first step of the EDP (Figure 1.1). The goal of this step is to understand the importance of the design challenge, through a client's motivations, and the desired outcome from the project, through a client's expectations. By clarifying the assignment, a team can reach consensus about what a problem is and why it needs to be solved. This information is used to write a clear problem statement.

Another key part of this step is distinguishing project features that need to be included for the user from attributes desired by the client. This task involves analyzing the assumptions of the client. For example, an engineering team may be tasked with designing a cycling wheelchair for an athlete with cerebral palsy. While the athlete (user) may prioritize the comfort of the seat, the client's priority could be using simple materials to minimize overall solution cost. In this step of the process, the engineer (designer) clarifies and balances these expectations, which will drive the solution design.

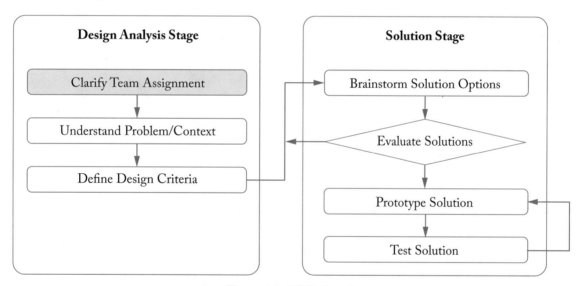

Figure 1.1: EDP: Step 1.

By the end of the section, you should be equipped to:

- Define the scope of a design problem,

- Recognize the importance of defining the problem to be solved,

- Differentiate between the involved parties of a project (user, client, designer),

- Effectively conduct a client interview, and

- Clearly state the problem to be solved.

Playlist

Watch the following video playlist to get an overview of the clarifying the team assignment step. Learn to distinguish between the key stakeholders of a project, as well as how to conduct a client interview.

| Videos | bit.ly/step01-clarifyingteamassignment |

Video Notes

Define the following key terms.

- User-designer-client triangle:

- Problem statement:

What are the key distinctions between the participants in the EDP?

Describe differences between motivations, assumptions, and expectations.

Content Summary

The purpose of clarifying the team assignment is to define the scope of a project. Often, this culminates in a written problem statement, which succinctly describes the problem that needs to be solved. Clear problem statements typically have a user, an action, and an outcome. Examples include:

- Individuals with disabilities need an easy and safe way to get around their community.

- Children with musculoskeletal disorders need a better way to feed themselves.

(Note that the examples state a need for a solution. "I want a new toothbrush" is not a problem statement.)

Before a designer can write a problem statement, they must identify the client's motivations, assumptions, and expectations. Client <u>motivations</u> to pursue a project are often varied. Some may be moved by a perceived need or desire to solve a problem. Others may seek a new technology or a business opportunity. Although the designer is responsible for actualizing a solution, clients often have <u>assumptions</u> about what the solution will look like in its end stages. They may assume a final design's features, cost, and aesthetics. Finally, <u>expectations</u> of the design team vary with each client. Some clients may have stringent objectives, constraints, or functions, while others may leave the designer with more freedom.

Teasing out a client's motivations, assumptions, and expectations is best achieved during an in-person interview. When brainstorming questions to ask during the client interview, keep in mind:

1. Open-ended questions allow the client to elaborate, and

2. The different points of the user-designer-client triangle (Figure 1.2).

The user-designer-client triangle (Figure 1.2) is a visual cue to remind the designers (engineers or engineering students) that users (the individuals who will use the product or design) are not the same as the client (the individual who brought the problem to the designer). In special cases, the client and the user may be the same person.

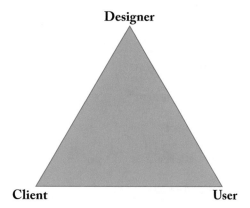

Figure 1.2: User-designer-client triangle.

Review Questions

1. Clarifying the team assignment is the _____ step of the EDP.

   A. 1st

   B. 2nd

   C. 3rd

   D. 4th

2. If you were tasked with developing a new syringe, who would be potential users?

   A. Your team

   B. A professor

   C. Nurses

   D. Medical device companies

3. If you were tasked with developing a new syringe, who would be the client?

   A. Your team

   B. A professor

   C. Nurses

   D. Medical device companies

4. Which of the following is true of the designer?

   A. The designer bears the primary responsibility for solving the engineering problem.

   B. The designer is sometimes the same as the client.

   C. The designer is often the same as the user.

   D. The designer sponsors and funds the project.

5. After completing the "Clarify team assignment" step, teams should ____.

   A. Understand the purpose of the project

   B. Understand the problem the project seeks to address

   C. Understand the client's desires and needs

   D. All of the above

6. A client's ____ for a project may include the absence of existing solutions, a demonstrated need for a device, new business opportunities, and the development of new technologies.

   A. Wishes

   B. Goals

   C. Motivation

   D. Potential ideas

Exercise #1

Introduction

Giraffes at the Zoo

There are numerous Masai giraffes that currently call the local zoo home. Giraffes are the tallest living animals; with long necks, they reach plants high off the ground (Figure 1.3). They have a rich diet of leaves, hay, shrubs, twigs, and even fruit. The Masai giraffes live primarily in Tanzania and southern Kenya, although other varieties are found throughout Africa.

Enrichment Devices Stimulate Lives of Animals

Keeping animals in captivity entails adhering to regulations imposed by the United States Department of Agriculture (USDA). For zookeepers, it is a responsibility to cultivate a stimulating life for all animals in the facility. This cultivation is achieved through enrichment activities, which are often devices that elicit natural behaviors and challenge the strength and mental agility of the species. Enrichment items for giraffes on exhibit at the zoo are currently limited.

Project Goal

The goal of this project is to create a hay feeder for giraffes at the zoo. This device should also be an enrichment activity to help mentally stimulate the giraffes on exhibit.

Desired Characteristics of Device

All additions to the zoo, including enrichment devices, must look natural and mimic the existing habitat. Ideally, the device should be placed at the front of the exhibit to draw the giraffes toward the visitors. The device must be sturdy enough to withstand interaction with the giraffes. The device should also be able to withstand all local weather conditions; it should be durable and last at least 3 years. The device should be fun for the giraffes and engage them on a regular basis. Cost is a consideration.

Figure 1.3: **Masai giraffe.**

Task

Identify possible expectations of the client for the giraffe project at the zoo. You can underline direct phrases in the project prompt.

Identify possible assumptions of the client. You can underline direct phrases in the project prompt.

After reading the prompt, the design team for the giraffe project summarized key objectives below. Using this information, generate at least five key questions for the client interview.

Objective 1: The device must be durable.

Objective 2: The device should be engaging for the giraffes.

Objective 3: The device should look natural.

Questions:

Exercise #2

Introduction

Limited Motor Function in Children

Musculoskeletal disorders may affect individuals by limiting their range of motion of their arms and hands, as well as their grip strengths. These limitations can have a great impact on everyday activities. Children may need assistance, as their weak coordination and motor skills can make seemingly easy tasks very difficult to complete independently.

Need for Independence with Eating

An individual with a musculoskeletal disorder may need help to cut food, transfer food from the plate to the utensil, raise the utensil with food to their mouth, and wipe their mouth afterward. When possible, therapists and parents support children so they can develop the ability and confidence to independently feed themselves. Although there are a variety of utensils, including spoons and forks, that have been proposed for children with other disabilities, few are sufficient for children with more extensive musculoskeletal disorders. The available utensils require children to have wide ranges of motion and require children to have enough gripping power to hold them.

Problem Statement

The goal of this project is to design a low-cost, comfortable, and efficient utensil for children with musculoskeletal disorders.

Desired Characteristics of Utensil

The device should be able to assist a child with limited arm control and weak gripping strength to transfer food effectively. The device should be durable and able to withstand daily use, as well as be lightweight and easy to grip. Cost is also an important consideration for this device.

Task

Critique the following client interview questions and make suggested improvements as necessary. Rewrite poorly constructed or closed-ended questions. What category fits the question most closely: motivations (M), expectations (E), or assumptions (A)?

Category: M, E, A

1. Do normal spoons or forks spill frequently?

2. Should the eating utensil be light?

3. Should the utensil be made of metal?

4. Should children be able to eat from the device quickly?

5. Do you expect us to manufacture the device?

6. Who will use the utensil?

Exercise #3

Task

Critique the following problem statements. Evaluate their strengths and weaknesses. Does each statement have an action and an outcome?

1. The zoo needs a way to release food to animals periodically throughout the day.

2. Students need an app to find tools and materials in their makerspace.

3. It is difficult to find IV lines on a patient at night.

4. Students want to see which parking spaces are available when driving to school.

Applying This Step to Your Project

Task: Orient to the Project

Learn about your project by (re)reading the project description or refreshing your memory. You should tentatively be able to briefly state the problem that needs to be solved.

Task: Prepare for the Interview

Generate a list of questions for your client interview on the page entitled Client Interview Questions. At a minimum, the following topics need to be covered.

- Details on what is driving the design request. Why is the design needed? What is the problem that needs to be solved?

- What is currently used? What are the limitations? If nothing is available, why?

- Who will use the new design?

- How will the design be used?

- What characteristics are important in a new design? What are necessary or mandatory characteristics? What are some desired features?

- Logistical questions, including contact information of project sponsor(s).

Confirm that the questions are inclusive of what is needed to complete the client interview. Type up or neatly rewrite the questions for the interview. Order the questions so that there is some logical flow to the conversation. If possible, make copies of the questions for all team members before the interview.

Task: Conduct the Interview

Conduct a professional interview. Remember to NOT ask the client for their "solution." Also, don't postulate your possible solution ideas during the interview! One or two (but not all) team members should take notes. This can be done on the page entitled Client Interview Notes. All members should be active listeners during the interview. The best practice is to open the interview by asking the client to restate the problem and close by asking for their contact information.

Task: Summarize the Project Goals

Reflect on the information obtained during the client interview. Your team should be able to clearly identify the problem and motivation, the user(s), and other key project information. Complete the page entitled Project Summary.

Client Interview Questions

1.

2.

3.

4.

5.

6.

7.

8.

9.

10.

11.

12.

13.

14.

15.

16.

17.

18.

19.

20.

Client Interview Notes

Project Summary

What problem are you trying to solve? What is the unmet need or unique opportunity?

Why is this problem important?

Who is the user? Who will be using the solution?

Who will be working on this project? Include project members and contact information.

Who is the client? Include contact information.

Project Summary

CHAPTER 2

Step 2: Understand the Problem and Context

Understanding the problem and context is the second step of the EDP (Figure 2.1). Having in-depth knowledge about the entire problem space allows teams to make research-based, justifiable decisions. This step involves producing a need-to-know list for a specific project and conducting a comprehensive literature review. For example, if an engineer was attempting to develop a giraffe feeder, they would need to know what a giraffe eats, how much a giraffe eats, and the method by which a giraffe eats. By recording this need-to-know list, the engineer can then turn to literature from zoologists who have studied giraffes to answer these questions.

It is important to complete this research in order to select meaningful design criteria and generate a breadth of brainstormed ideas. However, it should be noted that research is also iterative; an engineer may need to revisit and revise their need-to-know list regularly throughout the entire design process.

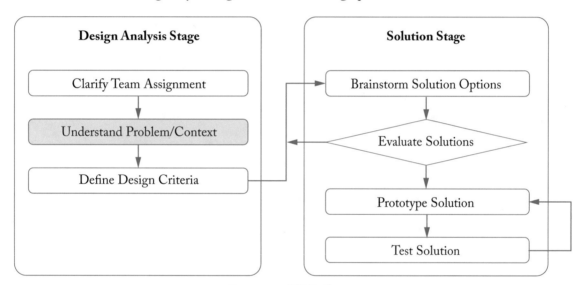

Figure 2.1: EDP: Step 2.

By the end of the section, you should be equipped to:

- Explain a method to organize and categorize the research foci for a problem space,

- Generate an in-depth need-to-know list for research beyond surface-level information,

- Conduct extensive research, focusing on reliable, peer-reviewed sources, and

- Summarize key findings from research.

Playlist

Watch the following video playlist to get an overview about understanding your design problem and its context. Learn how to categorize the problem space to obtain informative and relevant research.

Videos	bit.ly/Step02-understandingtheproblemandcontext

Video Notes

Define the following key terms.

- Need-to-know list:

- Problem context:

- Peer-reviewed:

What are the four subcategories of the problem space?

What is the key difference between a deep and a shallow dive during research?

List at least two examples in each of the subcategories for the water bottle project in Table 2.1.

Table 2.1: Water bottle project problem space	
Subcategory	Examples

Content Summary

The purpose of understanding the problem and context is to take a deep dive into the problem space, with the goal of fully understanding your problem. Your team needs to gather and study information relevant to your project. Figure 2.2 provides a range of topics to explore. Details on these topics are given in Table 2.2.

Your team may need to consult with your client again after completing the research. Make sure to utilize good research techniques, such as using peer-reviewed sources.

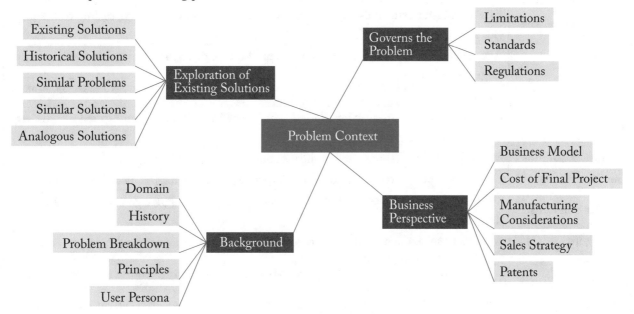

Figure 2.2: **Problem space.**

Table 2.2: Problem space breakdown	
Background	
Domain	Relevant definitions, terminology, and diagrams, e.g., define BPA and "clean water"
History	History of your design or similar designs, e.g., water bottles were made with glass in the 1800s
Problem Breakdown	Components of the problem that need to be researched, e.g., storing water, bottling liquids, designing containers
Physical Principles	Engineering fundamentals, equations, laws of physics, and graphs, e.g., sealing technologies and heat transfer fundamentals
User Persona	Age, gender, occupation, and needs of a typical user, e.g., any swimmer that needs an easy-to-grip water bottle
Exploration of Existing Solution	
Existing Solutions	Method or device currently being used, which may include active and expired patents, e.g., water bottles with novel shapes
Historical Solutions	Method or device that was used in the past but is no longer in use, e.g., glass water bottles
Similar Problems	Other situations with similar concerns, e.g., transporting liquids in closed container
Similar Solutions	Other methods or devices with functions similar to your design, e.g., texture on barbells that helps with grip
Analogous Solutions	A problem solved in a similar way that is outside of your domain, e.g., cereal boxes and hay feeders both hold and transport food
Governs the Problem	
Limitations	Reasons for the lack of a solution to your problems, e.g., market size, current technologies
Standards	Rules made by organizations over product performance and tasks, e.g., size of the threading on screw cap of water bottle
Regulations	Rules, usually made by the government or other organizations such as ASTM, ASME, IEEE, and NFPA that must be followed, e.g., FDA regulations on water quality for bottled water
Business Perspective	
Business Model	Is a business model needed for the design? e.g., a single coffee portioning device made for NASA may not need a business model
	Product distribution to users, e.g., an app could be downloaded remotely, a device could be shipped to customers or sold in stores
	Type of company, e.g., NGO, for profit, non-profit, none
Cost of Final Product	Factors that affect costs, such as manufacturing or packaging, e.g., cost of an extra $0.10 if water bottles are sold in cardboard boxes
Manufacturing Considerations	Materials and processes needed to produce on a larger scale, e.g., differences when make 1000 bottles in a factory
	Use of existing manufacturing streams and machines, e.g., applicability of current water bottle manufacturing machines
Sales Strategy	Market analysis: size of market ($ and people), target customers, e.g., reusable water bottles sales are annually $1.5 billion, over 330,000 people are members of USA Swimming
Patents	Active patents that exist on unique features of your design, e.g., patents on cooling/heating technology for containers

Review Questions

1. When researching the domain of a problem, your team should cover which of the following? Select all that apply.

    A. Definitions

    B. Previous solutions

    C. Client expectations

    D. Terminology

    E. Diagrams

2. ____ inform the design context review.

    A. People

    B. Documents

    C. Destinations

    D. All of the above

3. A shallow dive consists of ___. Select all that apply.

    A. Typing a key word into Google and selecting the first couple of links

    B. Citing peer-reviewed sources

    C. Not developing a need-to-know list that covers the breadth of the problem

    D. Using Wikipedia as a major source

4. Why is it important to research existing solutions?

    A. Existing solutions can illustrate weaknesses in existing designs.

    B. Your team can improve upon an idea found in an existing solution.

    C. Your team can understand what materials are available and practical to use.

    D. All of the above.

5. Which of the following are verified sources you could use in researching the problem and its context? Select all that apply.

    A. Textbooks

    B. Review papers

    C. Company websites

    D. Government documents

Exercise #1

Introduction

A Universal Tool for Carrying Materials

Backpacks are a great invention for holding, storing, and transporting all kinds of objects (Figure 2.3). There are numerous styles of backpacks for any activity—from hiking to kayaking, shopping to school. Backpacks can differ in their storage capacity, shape, interior features, and overall aesthetic. In today's consumer market, backpacks can be customized to meet the needs of a range of clients.

The Need for an All-Purpose Packing Solution

Even though backpacks can be highly customized, they still tend to fall in a category depending on their function. For instance, while everyday backpacks can differ in color and design, they are still generally not able to support large weight. There is a need for a new backpack design that can be used for everyday wear and accommodate heavier loads when camping or participating in other outdoor activities.

Project Goal

The goal of this project is to design a dual-purpose backpack for hiking and everyday wear that is lightweight, aesthetically pleasing, and spacious.

Desired Characteristics of Backpack

The backpack should be durable and able to carry at least 25 lb of weight. It

Figure 2.3: Backpack.

should also be weather-resistant for use outdoors. The backpack should be comfortable to wear for long periods of time. The backpack should be spacious and lightweight when not loaded with materials. The backpack should be aesthetically pleasing.

Task

Create a need-to-know list for the backpack project. When applicable, highlight words directly from the project prompt. Be sure to account for all subcategories of the problem space.

Background

Exploration of Existing Solutions

Governs the Problem

Business Perspective

Exercise #2

Introduction

Risk of Severe Illness in Newborn Children

More than 60% of newborns have a condition called neonatal jaundice, which is the yellowing of the skin. The cause of this is the accumulation of the orange-yellow substance bilirubin produced naturally through the breakdown of red blood cells. Neonatal jaundice often clears up without medical intervention; however, in severe cases, high levels of bilirubin can lead to brain damage. Neonatal jaundice is frequently treated by exposing the child to blue light, called phototherapy (Figure 2.4). One new, low-cost phototherapy option shown to work in Malawi deploys a cassette of 470 nm LEDs.

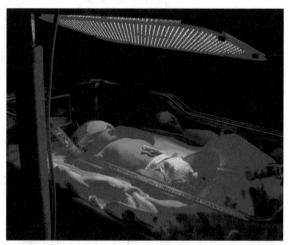

Figure 2.4: Infant under phototherapy lights.

Delivery of Phototherapy Lights Needs More Flexibility

The cassette of phototherapy lights was initially designed to work with a baby incubator. In this designed configuration, a fixed height between the baby and the lights was established that guaranteed the right intensity of blue light be delivered to the baby. However, incubators are not widely in use in sub-Saharan Africa, and babies are placed on all types of surfaces. This is especially true in low-income communities, such as Malawi. Without an easy or convenient method to set the phototherapy lights above the baby at the correct height, some babies are not receiving the proper dose of light as treatment.

Project Goal

The goal of this project is to develop a low-cost, flexible device for positioning phototherapy lights for the babies at Queen Elizabeth Central Hospital (QECH) in Blantyre, Malawi.

Objectives and Constraints

The device should be able to adjust the lateral position and height of the lights relative to the position of the baby. The device must be flexible enough to work for different physical scenarios, from cribs and incubators to flat surfaces. The device should be durable and made from materials available in Malawi. The device should be easy to use. The device must be low cost and simple in its design.

Task

An engineering design team who is developing a device for positioning phototherapy lights generated the following need-to-know list after the client interview.

- <u>Overview of jaundice and phototherapy</u>

  - The causes and effects of jaundice

  - A brief outline of treatments for jaundice other than phototherapy

- <u>Current phototherapy practices in QECH</u>

  - The types of phototherapy lights used in QECH

  - Phototherapy protocols

    - Proper distances that the lights should be held from the patient

- <u>Ministry of Health</u>

  - How the Ministry of Health is involved in phototherapy

    - Regulations, laws, protocols, standards, oversight

- <u>Methods of phototherapy delivery</u>

  - Current methods of holding and supporting phototherapy lights

  - Technologies that are similar in function to supporting and holding phototherapy lights

- <u>Manufacturability</u>

  - Who would construct the potential device

  - How the potential device would be constructed

  - The cost constraints of the potential device

  - Who would buy the potential device

Critique this need-to-know list. What did the team do well? Describe any topics the team is missing.

Applying This Step to Your Project

Task: Create a Need-to-Know List

Your team should generate a need-to-know list that suggests topics and issues to research based on Figure 2.2. For each main area, consider the listed categories or topics of research (e.g., historical solutions or regulations). List two or three specific research topics related to your design project. These topics should be keywords that could form the basis for an initial literature search.

Task: Complete Research

Once a complete need-to-know list for all four areas has been generated, break up the research among your team members and assign work. Query verified or peer-reviewed resources to uncover knowledge about your project.

The "Exploration of Existing Solutions" section is emphasized because all teams must have strong knowledge in this area. Some topics for possible exploration include the following.

- Descriptions of current technologies and existing solutions that inadequately address or attempt to solve your design challenge.

- Descriptions of solutions to similar problems.

For some projects, learning about the Background is critical, whereas for others, applicable regulations are key. For most projects, the topics under Business Perspective can be deemphasized until further along in the prototyping process. Since the most critical research topics vary among projects, talk with an instructor or advisor to identify the focus of your research. It is likely that you will have research topics across the different quadrants; avoid just doing the "easiest." Some topics for possible exploration include the following.

- Specific domain knowledge (e.g., physiology of the disease area or process with which your project is associated).

- Applicable theories, principles, or technical approaches/methodologies that apply to the implementation or design of your problem or possible solutions.

- Potential users of your design (their preferences, demographics, etc.).

- Limitations, standards, and regulations governing the design.

- Any environmental or cultural issues that affect how your design will be used and implemented.

- Existing patents, business models, or sales strategies.

Conduct your assigned research, using the research reference worksheets on the subsequent pages. You may need more than the five pages provided. Be sure to include enough information on the source that the reference can be returned to later. The summary of the reference should be thorough.

Research Reference Worksheet

Problem Context Area (circle one that best fits):

    Exploration of existing solutions   Governs the problem   Background   Business perspective

---

Complete Source:

    Is this source peer-reviewed (circle one):   Yes   No

Summary of Key Points:

Why is this source important to the project?

Problem Context Area (circle one that best fits):

    Exploration of existing solutions   Governs the problem   Background   Business perspective

---

Complete Source:

    Is this source peer-reviewed (circle one):   Yes   No

Summary of Key Points:

Why is this source important to the project?

Research Reference Worksheet

Problem Context Area (circle one that best fits):

Exploration of existing solutions   Governs the problem   Background   Business perspective

Complete Source:

Is this source peer-reviewed (circle one):   Yes   No

Summary of Key Points:

Why is this source important to the project?

Problem Context Area (circle one that best fits):

Exploration of existing solutions   Governs the problem   Background   Business perspective

Complete Source:

Is this source peer-reviewed (circle one):   Yes   No

Summary of Key Points:

Why is this source important to the project?

Research Reference Worksheet

Problem Context Area (circle one that best fits):

Exploration of existing solutions   Governs the problem   Background   Business perspective

Complete Source:

Is this source peer-reviewed (circle one):   Yes   No

Summary of Key Points:

Why is this source important to the project?

Problem Context Area (circle one that best fits):

Exploration of existing solutions   Governs the problem   Background   Business perspective

Complete Source:

Is this source peer-reviewed (circle one):   Yes   No

Summary of Key Points:

Why is this source important to the project?

Research Reference Worksheet

Problem Context Area (circle one that best fits):

    Exploration of existing solutions   Governs the problem   Background   Business perspective

Complete Source:

    Is this source peer-reviewed (circle one):   Yes   No

Summary of Key Points:

Why is this source important to the project?

Problem Context Area (circle one that best fits):

    Exploration of existing solutions   Governs the problem   Background   Business perspective

Complete Source:

    Is this source peer-reviewed (circle one):   Yes   No

Summary of Key Points:

Why is this source important to the project?

Research Reference Worksheet

Problem Context Area (circle one that best fits):
Exploration of existing solutions   Governs the problem   Background   Business perspective

Complete Source:
Is this source peer-reviewed (circle one):   Yes   No

Summary of Key Points:

Why is this source important to the project?

Problem Context Area (circle one that best fits):
Exploration of existing solutions   Governs the problem   Background   Business perspective

Complete Source:
Is this source peer-reviewed (circle one):   Yes   No

Summary of Key Points:

Why is this source important to the project?

CHAPTER 3

Step 3: Define Design Criteria

Establishing design criteria is the third step of the EDP and the final step of the design analysis stage (Figure 3.1). All projects need targets and goals that describe features and characteristics of the design. These goals drive design decisions as the solution is developed. For example, if a team is tasked with creating a backpack that should cost less than $10, then this target design criteria guides the selection of the solution materials, features, and overall shape. Whether the design meets the established criteria is evaluated and tested as the prototype evolves toward a final solution. Establishing reasonable, relevant design criteria is critical—as these target values serve as a guidepost throughout the remainder of the design process.

Three substeps to developing a refined list of design criteria are discussed in this workbook. In Step 3A: Establishing Objectives and Constraints, there is an explicit discussion of design objectives and constraints—as well as how to quantify these as design criteria by defining them with a number.

When defining design criteria, you may discover that some are more difficult to quantify as they depend on feelings and opinions, which differ from person to person. The content in Step 3B: User-Defined Scales describes a user-defined scale, which is a tool for quantifying criteria that rely on feelings and opinions.

In preparing your list of design criteria, it may become clear that some design objectives are in opposition (e.g., low cost and high durability), or that some design objectives are more important than others. A pairwise comparison chart (PCC) is a graphical tool that helps prioritize design objectives. Step 3C: Pairwise Comparison Chart describes the steps to create a PCC.

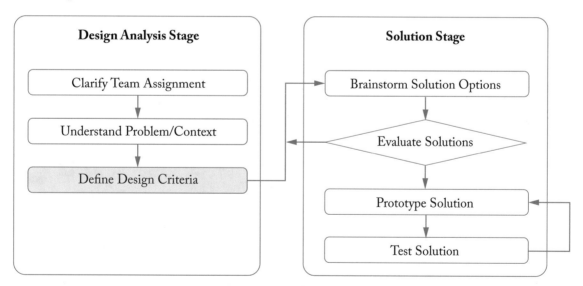

Figure 3.1: EDP: Step 3.

By the end of the section, you should be equipped to:

- Differentiate between design objectives and constraints,

- Quantify design criteria for a project,

- Evaluate qualitative criteria through user-defined scales, and

- Rank design criteria through a pairwise comparison chart.

STEP 3A: ESTABLISHING OBJECTIVES AND CONSTRAINTS

Playlist

Watch the following video playlist to get an overview about design objectives and constraints. Learn how design criteria are created from objectives and constraints by setting a numerical target value.

Videos	bit.ly/step3a-establishingobjectivesandconstraints

Video Notes

Define the following key terms.

- Design criteria:

- Objectives:

- Constraints:

- Functions:

- Means:

What are the differences between objectives and constraints?

What are the differences between functions and means?

Why is it important that design criteria are quantitative?

Content Summary

Design criteria are used to quantitatively describe design objectives and constraints, either with a single number or a range of values. Objectives are desired attributes, while constraints are strict limits a design must meet. Constraints are binary and must be satisfied or the design will fail. Most short-term design projects will have about two constraints and four to seven objectives. It is important to decide which criteria are critical for a design's success and which are only desired. Examples of measurements that can be made quantitative are listed below.

Number	Height	Electric resistance
Volume	Width	Conductance
Percentage of value	Angle	Magnetic field strength
Error of value	Frequency	Inductance
Concentration	Force	Curvature
Flow rate	Pressure	Pitch
Density	Energy	Amplitude
Repeatability	Heat	Slope
Cost	Charge	Accuracy
Power	Capacitance	Precision
Voltage	Resistance	

Remember that assigning a range of values is usually preferred, because exact values can be unnecessarily constraining. When possible, use >, <, between, or within to define your value. Avoid using =, as it is often hard to get something exactly!

Review Questions

1. The purpose of establishing design criteria is to allow the design team to ____.

   A. Define and quantify several key objectives that should be met in the final solution

   B. Determine the best solution of all brainstormed ideas

   C. Prioritize the key objectives of the final design

   D. Measure whether or not the design solution is meeting its target goals

2. Which of the following are acceptable design criteria? Select all that apply.

   A. Lasts >5 years

   B. Flow rate of >1 gallon/minute

   C. Affordable

   D. Complies with applicable FDA regulations

3. Which of the following are examples of measures that can be made quantitative? Select all that apply.

   A. Flow rate

   B. Energy

   C. Texture

   D. Pressure

   E. Repeatability

4. Design criteria are defined as ____.

   A. Things so important their absence means failure

   B. Measures that can be made quantitative such as numbers or bounded ranges

   C. The designers' own experiences with the key problem being solved

   D. Qualitative descriptions of the desired attributes

5. Which of the following types of boundary conditions are preferable when defining design criteria? Select all that apply.

   A. > or <

   B. =

   C. Between

   D. Within ± %

6. Your client has stated that she would like a design for bike wheels for low-income settings to be low cost. Your research shows that a typical wheel in the United States costs $50. Which of the following is a reasonable design criterion for your team?

   A. $50–$70

   B. ~$50

   C. =$25

   D. <$8

Exercise #1

Introduction

Your design team is tasked with building an improved hay feeder for the giraffes at your local zoo. In talking with the keepers, you identify three design objectives: safety, food capacity, and durability. You interview the keepers and conduct research. Here is a transcript of an interview with the keeper, along with related research on each of the criteria.

Team member: What would make a hay feeder safe and unsafe to the giraffe herd?

Keeper: The feeder cannot be composed of materials that may pose a threat to the giraffes (e.g., loose small parts, toxic paints, sharp edges). Also, the design must prevent the ossicones from getting stuck in the feeder.

Research: Ossicones are the small horns on the top of a giraffe's head (Figure 3.2). Ossicones are typically two inches in length. Giraffes also have long tongues, typically 18 inches in length. Giraffes regularly eat leaves, hay, fruits, and vegetables (including whole carrots).

Figure 3.2: **Masai giraffe**

Team member: How much food needs to be placed in the feeder at once?

Keeper: We would like a feeder that can be filled once each day. About half of the food consumed by giraffes in captivity is hay. For each giraffe, we typically load five flakes of hay per day into the feeder. This amount is limited by what we can easily carry.

Research: Each giraffe eats 30–75 lb of "food" (leaves, hay, shrubs, twigs, fruit) daily. The hay constitutes half of their daily food intake at the zoo. One bale of hay weighs ~ 60 lb and there are ~12 flakes of hay per bale. One flake of hay is 3–4 inches by 13–15 inches by 10–12 inches.

Team member: What about durability?

Keeper: It would be great if it lasted a year. We would prefer to have minimal maintenance on the device, maybe less than once per month.

Task

Write quantitative design criteria for food capacity and durability.

Exercise #2

Introduction

An engineering team was tasked with designing a new utensil for children with musculoskeletal disorders. The team wrote the following design criteria for the five identified design objectives.

Task

Critique the listed design criteria and make suggested improvements as necessary. Rewrite poorly constructed design criteria.

- Reduced feeding time: Feeding time is reduced by 50% of current feeding time.

- Transport: 80% of the food is transferred to the child's mouth.

- Ease of use: >70% of the children can grip the spoon and use it.

- Lightweight: >200 g

- Cost: <20

Applying This Step to Your Project

Task: Identify Design Objectives and Constraints

Review the interview notes from your client and/or potential users as well as the initial project description. Also, recall the research that you completed in the previous step of the design process. From these sources, work with your team to generate a list of design objectives and constraints. Your team may also list functions and means.

On Table 3.1, start by listing out the objectives and constraints without any numbers. Examples of these objectives and constraints are weight, cost, and ease of use. Identify if any objectives are really constraints, meaning that they must be present for a solution to be successful. Confirm a reasonable list of design constraints (typically 1–2) and objectives (typically 4–7).

Task: Quantify Design Criteria

After there is consensus on this list, quantify each design objective and constraint. To determine appropriate numerical values, refer to the project statement, conversations with the client, and research worksheets. Avoid making up numbers. Note that you may need to consult with your client again if you don't have all the information that you need. Make sure that your design criteria can be measured or evaluated in a quantitative way.

If any of your design constraints or objectives are feelings or opinions, you may need to establish user-defined scales. User-defined scales are described in Step 3B: User-Defined Scales.

Task: Reflect on Design Criteria

Consider the following.

- Do these numbers make sense? Are these numbers within a range that is appropriate for a design project?

- Do any of these include feelings or opinions? If so, go to Step 3B: User-Defined Scales.

- It may be important for you to rank the importance of some of these design criteria. If so, go to Step 3C: Pairwise Comparison Chart to fill out a PCC for your design criteria.

Table 3.1: Design criteria

Criterion	Target Value	Justification

Objectives   Constraints

STEP 3B: USER-DEFINED SCALES

Playlist

Watch the video playlist to learn more about user-defined scales (UDS), which are used to quantify feelings and opinions.

Videos	bit.ly/step3b-user-definedscales

Video Notes

What is the difference between qualitative and quantitative design criteria?

What are some examples of qualitative design criteria?

What is a level in a user-defined scale?

What is an example of a target criterion for a user-defined scale?

Content Summary

Your project's goals should be measurable, meaning you should be able to prove when they are met or completed. Sometimes when you review your objectives, you find that they might be a bit vague. For example, you may hear your client describe their needs using some of the following qualitative words.

Appearance	Ease of use	Taste
Sound	Color	Smell
Safety	Texture	Comfort

It is difficult to associate a number with this type of data as it is based on personal opinions or feelings about the characteristics of objects. Also, this information can change based on person or time.

User-defined scales (UDSs) are a formal way for design teams to convert qualitative objectives to quantitative design criteria. Table 3.2 is an example of a UDS for how easy or difficult it is to drink from a water bottle. This scale assigns a score of 1–3 based on an individual opinion in using the water bottle. Figure 3.3 shows a Likert scale, which teams use to test a user's agreement with a statement. For example, a user could respond with "Agree" for the statement "It is easy to drink from this water bottle."

Table 3.2: User-defined scale for new water bottle design	
Level	**Attribute**
1	Easy to use
2	Neither easy nor difficult to use; neutral
3	Difficult to use

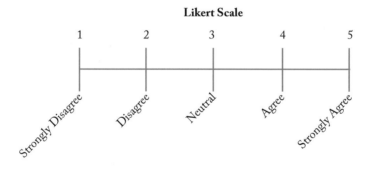

Figure 3.3: User-defined scale for agreement with statement.

Review Questions

1. In order to define the design criterion for a UDS, the team must _____.

    A. Identify the constraints of the design

    B. Survey at least ten users

    C. Select a numerical value or range of values that must be reached

    D. Discuss the importance of that objective with the users

2. When creating a UDS, the design team should set a number of levels. Which of the following is NOT recommended for the number of levels within a scale?

   A. 3

   B. 5

   C. 10

   D. 20

3. The Likert scale is an example of a widely used UDS for surveying people. What does it measure?

   A. Likelihood of a potential user buying the product being tested

   B. Agreement with a statement

   C. Quality of a product compared to a competitor's product

   D. Attitude toward the subject of the survey

4. Which of the following are steps involved with defining a design criterion for a qualitative objective? Select all that apply.

   A. Assign specific attributes to each level of the scale.

   B. Select number of levels in scale.

   C. Collect data of the responses from users to see if the criterion is met.

   D. Set the score that must be achieved for the design criterion to be met.

5. What should the attribute be for level 4?

Level	Blends into exhibit
5	Does not notice even when told the location of the device
4	?
3	Notices device once told to look for it
2	Notices device after viewing the exhibit for some time
1	Notices device immediately

   A. Notices device within 1 minute

   B. Notices device within 5 minutes

   C. Does not notice device for some time

   D. Notices device once told the location

6. What is wrong with this UDS?

Level	Easy to Use
5	Takes <2 seconds to remove lid
4	Takes 3–5 seconds to remove lid
3	Takes 5–10 seconds to remove lid
2	Takes 10–20 seconds to remove lid
1	Takes >20 seconds to remove lid

A. It has uneven intervals

B. It is unnecessary

C. It has too many levels

D. It has too few levels

Exercise #1

Introduction

You have been tasked with developing a new spill-proof coffee mug (Figure 3.4). After speaking with the client and ten potential users, you identified five design objectives. One of them was "ease of cleaning." Your team has created the following UDS for this criterion.

Figure 3.4: Spill-proof coffee mug.

Ranking	Criterion: Ease of Cleaning
5	Requires less than 30 seconds to clean
4	Requires 30–45 seconds to clean
3	Requires 45–60 seconds to clean
2	Requires 60–120 seconds to clean
1	Requires more than 120 seconds to clean

Task

Identify the advantages and disadvantages of this UDS.

Generate two different UDSs that could be used to define "ease of cleaning" for this coffee mug.

For both UDSs, set a target value that must be reached for the criterion to be met.

Exercise #2

Introduction

You have been tasked with developing an umbrella (Figure 3.5) that is designed for an individual who is using crutches. After speaking with the client and ten potential users, you identified five design objectives. One of them was "easy to use."

Task

Write two or three questions that you would ask the client or users to further clarify this objective.

Figure 3.5: Umbrella.

List three different possible meanings for "easy to use" in the context of the umbrella design.

Create a scale to quantify "easy to use" for one of the meanings you identified in the question above. Be sure to assign attributes to each level of the scale.

Set the target value that must be met in order for the criterion to be satisfied.

Applying This Step to Your Project

Task: Develop User-Defined Scales

For design objectives and constraints that are not easily quantifiable, your team needs to develop UDSs. Refer to the project statement, conversations with the client and/or users, and research worksheets to gather information regarding UDSs. Follow these steps to develop any needed UDSs to quantify feelings or opinions.

1. Determine the number of levels in the scale.

2. Write descriptions that correspond with each level of the selected scale (e.g., 1–5).

3. Set a target value or range.

4. Record any lists or charts created.

Use the pages below to define and describe the UDSs.

User-Defined Scales

Design criterion:

Level	Attribute

Target value(s) desired:

Number of people to survey:

How this UDS may be administered:

User-Defined Scale

Design criterion:

Level	Attribute

Target value(s) desired:

Number of people to survey:

How this UDS may be administered:

STEP 3C: PAIRWISE COMPARISON CHART

Playlist

Watch the following video to get an overview about how to rank your design objectives using a pairwise comparison chart (PCC).

Videos	bit.ly/step3c-pairwisecomparisonchart

Video Notes

What is the purpose of a PCC?

What are common examples of constraints? How are constraints treated in a PCC?

What are the five key steps of completing a PCC?

What are common problems for teams when completing a PCC? List two examples and how they can be solved.

Content Summary

The purpose of the pairwise comparison chart is to prioritize the design objectives (Table 3.4). Your team should use the following steps to create a PCC.

1. List the objectives along the top and left-hand side.

2. Place dashes on the diagonal.

3. Compare the objectives.

   A. Write 1 in the row of the more important objective.

   B. Write 0 in the row of the less important objective.

4. Add the scores along each row.

5. Rank objectives and reflect.

Table 3.4: Example of a PCC	Ease of Use	Maintains Temperature	Durable	Cost	Aesthetics	Total
Ease of Use	---	1	1	1	1	4
Maintains Temperature	0	---	0	0	0	0
Durable	0	1	---	1	1	3
Cost	0	1	0	---	1	2
Aesthetics	0	1	0	0	---	1

Consider the following when completing a PCC.

- Internal consistency. For example, criterion A is more important than criterion B, and criterion B is more important than criterion C. Therefore, criterion A must be more important than criterion C.

- Embedded constraints. A constraint will always be more important than a design objective, so it will always win. Comparing two constraints for their relative importance cannot easily be done.

- Team disagreement. If the team has differences or disagreements about the priority of different criteria, you may need to consult the client.

Review Questions

1. The purpose of a PCC is to ____.

   A. Prioritize objectives

   B. Determine the goal of the project

   C. Eliminate objectives

   D. Identify the most challenging component of the project

2. A score of _____ signifies that one objective is lower in importance than the other objective.

   A. 0

   B. -1

   C. A dash

   D. 1

3. What does a three-way tie likely indicate?

   A. Lack of internal consistencies

   B. Disagreement among the team

   C. Embedded constraint

   D. Real tie

4. How do embedded constraints affect the results of the PCC?

   A. Constraints create outlier scores.

   B. Constraints always lose to design objectives.

   C. Constraints always win over design objectives.

   D. Constraints and design objectives are interchangeable, therefore they have no effect.

5. How do you ensure that there is an actual tie?

   A. Have a discussion with your team members.

   B. Have a discussion with your user/client.

   C. Ties cannot exist in a PCC.

   D. Both A and B.

6. Once complete, the PCC _____.

   A. Does not need to be used again

   B. Should give guidance in the evaluation of possible solutions

   C. Is given to the client

   D. Will be used for inspiration during the brainstorming process

7. According to the chart, which of the criteria is the most important?

	Cost	Portable	Useful	Durable	Total
Cost	----	0	0	1	
Portable	1	----	1	1	
Useful	1	0	----	1	
Durable	0	0	0	----	

A. Portable

B. Cost

C. Useful

D. Durable

8. If you want to show that portable is more important than cost, in which position should you mark a "1" on the chart?

	Cost	Portable	Useful	Durable	Total
Cost	A	B			
Portable	C	D			
Useful					
Durable					

A. Position A

B. Position B

C. Position C

D. Position D

Exercise #1

Introduction

An engineering team was tasked with designing and building an improved hay feeder for the giraffes at their local zoo. The team completed the PCC in Table 3.5 based on the following five criteria: safety, feeding time, food capacity, durability, and aesthetics.

Table 3.5: Pairwise comparison chart for giraffe feeder					
	Safety	**Feeding time**	**Food capacity**	**Durability**	**Aesthetics**
Safety	---	1	1	1	1
Feeding time	0	---	1	0	1
Food capacity	0	0	---	1	1
Durability	0	1	0	---	1/2
Aesthetics	0	0	0	1/2	---

Task

Based on this PCC, rank the design objectives.

Identify the internal inconsistencies in the PCC. Focus on feeding time, food capacity, and durability. Update the chart to be correct given that durability is of lower importance than feeding time.

Again, rank the objectives and confirm internal consistency.

Exercise #2

Introduction

Your design team is tasked with creating a new utensil for children with musculoskeletal disorders. After talking to an occupational therapist, your team has identified the following six design criteria.

Reduced feeding time	Lightweight
Transport	Durable
Ease of use	Cost

Below is a transcript of an interview with an occupational therapist (OT).

Team member: What motivated you to pursue this project?
OT: The project was motivated by the desire to get the children to eat independently and build self-confidence. Part of building self-confidence is also making sure the kids have time to interact with other children. Right now they spend so much time eating that they don't have as much time to play with others.

Team member: How important is the weight of the device?
OT: We consider the weight to be crucial. If the device is too heavy, the children won't be able to use it continuously during the lunch period. In a way, the weight of the device affects its ease of use.

Team member: Which is more important—transport of all the food from the plate to the mouth or reduced feeding time?
OT: If I had to choose one, I would consider effective transport of the food to be more important. The kids should be eating almost all of the food that their parents pack.

Team member: What about durability?
OT: It would be ideal if the device could be durable as well, especially because parents may not be able to fix the device once it breaks. However, I don't consider it a top concern.

Team member: Do you believe that cost should be a large factor in this design?
OT: It would really depend on the final design. If the device was very durable, I believe the parents would be willing to pay a higher price.

Task

Complete a PCC in Table 3.6 using the six design objectives.

Table 3.6: Pairwise comparison chart for eating utensil							
	DC1:	DC2:	DC3:	DC4:	DC5:	DC6:	Total
DC1:							
DC2:							
DC3:							
DC4:							
DC5:							
DC6:							

Based on the interview, what is the ranking of the six objectives?

What questions are missing from the interview?

Applying This Step to Your Project

Task: Complete a PCC

Complete a PCC (Table 3.7) for your team's design objectives. If your team has significant difficulty with ranking the objectives, consider contacting your project sponsor. Remember that design constraints should not be included in a PCC.

Table 3.7: Blank PCC

	DC1:	DC2:	DC3:	DC4:	DC5:	Total	Rank
DC1:							
DC2:							
DC3:							
DC4:							
DC5:							

CHAPTER 4

Step 4: Brainstorm Solution Options

Brainstorming is the fourth step of the EDP and the beginning of the solution stage (Figure 4.1). The goal of brainstorming is to generate as many ideas as possible to fully encapsulate the solution space. If an engineering team begins pursuing one solution without exhausting all other possibilities, they may overlook an idea that fully satisfies all their design criteria. It is important to build a collaborative space where team members feel comfortable presenting out-of-the-box ideas and are willing to solicit critical feedback about their ideas. This will ensure that a wide range of ideas are considered.

Two substeps for effective brainstorming are discussed in this workbook. In order to generate ideas for all aspects of a solution, it is often helpful to decompose a project into smaller, independent parts. This process of decomposition is outlined in Step 4A: Decomposition. Then, various tools and methods for generating ideas are described in Step 4B: Brainstorming.

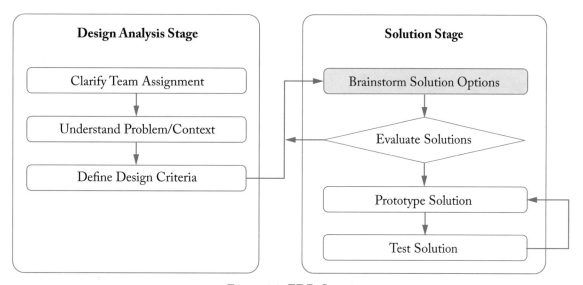

Figure 4.1: **EDP: Step 4.**

By the end of the section, you should be equipped to:

- Identify the key functions of a design solution,

- Decompose a project into discrete design blocks,

- Brainstorm many novel ideas to solve a problem,

- Apply two or more strategies to generate a breadth of ideas, and

- Sort solution ideas into relevant design blocks.

STEP 4A: DECOMPOSITION

Playlist

Watch this video playlist for an overview about decomposition. Learn how to naturally break a problem or project into many different, discrete parts.

Videos	bit.ly/step04a-decomposition

Video Notes

Define the following key terms.

- Decomposition:

- Design blocks:

What types of projects benefit from decomposition?

What are two key characteristics of design blocks?

What are the benefits of using decomposition within a team? What must a team be careful of during this process?

Content Summary

Decomposition is a process in which a large design problem is separated into smaller, manageable, independent chunks. The smaller parts are called design blocks. These blocks can be created based on functions, features, or physical aspects. Recall, functions are actions that a solution performs. Features are desired attributes for a solution. Physical aspects are discrete pieces or parts of an expected design solution.

The figure below provides an example of a decomposed project. Figure 4.2 decomposes a bicycle by its *physical aspects*. The device has three primary mechanisms, one for driving, one for control, and one for providing structural support to the rider.

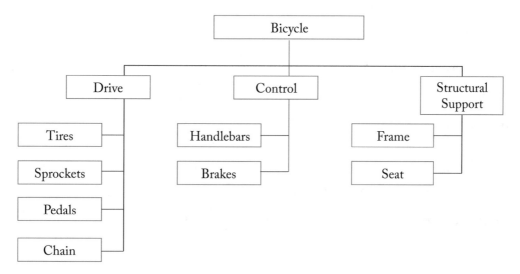

Figure 4.2: Example decomposition for a bicycle.

Review Questions

1. Decomposition is a process that involves ____.

   A. Separating a large design problem into smaller chunks

   B. Sorting through all of the brainstormed solutions and selecting the most promising ideas

   C. Brainstorming new ideas

   D. Ranking solutions based on faithfulness to design criteria

2. Decomposition succeeds when ____.

   A. The Card Method is used

   B. Individuals on the team are arbitrarily assigned parts of the project

   C. The team brainstorms several solutions

   D. Ideas are separated into discrete project blocks

3. Which of the following are the two most important considerations in selecting project blocks? Select two answers.

   A. Ability to be worked on independently from one another.

   B. Capability of the team to subdivide work into blocks that can be accomplished by specific members of the team.

   C. Ability of an individual to complete a part of the work.

   D. Cohesiveness with the overall design solution.

4. Decomposition can be helpful for _____.

   A. Brainstorming solution options

   B. Sorting ideas into different decision matrices

   C. Prototyping

   D. All of the above

5. Decomposition during brainstorming _____.

   A. Leads to an infinite number of ideas and slows a team down

   B. Helps a team focus on aspects or features

   C. Is generally not effective

   D. Generates fully formed solution ideas

6. Decomposed sections of a project _____.

   A. Can be used in a morph chart

   B. Tell a team whether a certain design block is necessary

   C. Are helpful when performing research

   D. Should immediately be sent to the client

Exercise #1

Introduction

You have been tasked with decomposing a refrigerator into discrete design blocks. These blocks should be relatively independent so they can be assigned to individual members of your design team.

Task

List at least four functions of a refrigerator.

List at least three features of a refrigerator.

Complete Table 4.1. Identify major design blocks and their associated functions and features. Think of what skills an individual team member may need to tackle each design block.

Table 4.1: Refrigerator design blocks		
Design Block	**Description (Functions and Features)**	**Needed Skill Set**

Exercise #2

Introduction

An engineering team was tasked with creating a dual-purpose backpack for hiking and everyday wear (Figure 4.4). The team wrote down all the components of the backpack and sorted them into design blocks.

Task

Identify what functions correspond to each of the following design blocks for the backpack.

- Pockets:

- Backstraps:

- Materials:

- Cushions:

Can any of these design blocks be further decomposed? If so, complete the decomposition.

Are there any design blocks missing? If so, list them and their corresponding functions.

Figure 4.4: Backpack.

Applying This Step to Your Design

Task: Complete Decomposition

Most, but not all, projects benefit from decomposition. Work as a group to decompose your design project into discrete design blocks. The design blocks can be separated by function, feature, or physical aspects. A good first step in decomposition is to list out the *verbs* associated with a potential solution. Consider the following questions.

1. What actions must the solution perform?

2. How will the user interact with a potential solution?

3. Does this solution involve movement, force, or a reaction?

4. What functions require multiple steps?

A next step in decomposition is to list features or physical aspects associated with potential solutions. Consider the following questions.

1. What parts might a solution have?

2. What desirable features might a solution have?

3. Is the solution contained in some way?

4. Are there parts that do or do not interact with the user?

List out the different design blocks with thorough descriptions. Continue to decompose until your team believes it will be easy to generate ideas for each block. Check with an instructor or advisor to confirm that the decomposed design blocks are reasonable.

STEP 4B: BRAINSTORMING

Playlist

Watch this video playlist for an overview about brainstorming. Learn how to generate many solution ideas with different, structured brainstorming methods.

Videos	bit.ly/step04b-brainstorming

Video Notes

What are two principles that govern brainstorming?

Why is it important to generate many ideas when brainstorming? How many ideas should you aim to brainstorm?

What is hitchhiking?

What should a team do to prepare themselves for brainstorming?

Content Summary

The purpose of brainstorming is to generate as many ideas as possible. When brainstorming, it is important to keep in mind these following rules.

1. Quantity over quality. The goal is to generate as many ideas as possible. With many ideas, there are likely several great ideas.

2. Wild ideas are welcome. Don't constrain yourself to obvious solution ideas. Generate crazy ideas. Often you can build upon these wild ideas to generate great ideas.

3. Hitchhiking is encouraged. When discussing ideas, build upon them to generate new ideas. Ideas belong to the group, not an individual.

4. Criticism is not allowed. Do not criticize other's ideas, no matter their quality. Do not self-sensor your own ideas; share them with the group.

The two methods of brainstorming mentioned in the videos are summarized in Table 4.2.

Table 4.2: Brainstorming methods	
Writing Slip Method	**Card Method**
1. Gather cards	1. Gather cards
2. Write/sketch one idea per card	2. Write/sketch one idea per card
3. Discuss ideas one at a time	3. Pass cards around team
4. Write new ideas on separate cards	4. Hitchhike and write new ideas on original card

One way to generate a variety of brainstormed solutions is to create out-of-the-box ideas. The SCAMPER method can help with rapid idea generation. The SCAMPER method is described below. Apply the "question words" to solution ideas to create more ideas.

Substitute

Who else instead? What else? Other place? Other time? Other process? Other power source? Other approach? Other tone of voice?

Combine

Blend? Combine purposes? Combine ideas? Combine units? Combine functions?

Adapt

What else is like this? What other ideas does that suggest? Ideas from the past to copy/modify?

Modify

Change meaning, color, motion, sound, appearance? New twist? What to add? Greater frequency? Stronger? Larger? Higher? Thicker?

Put to other uses

New ways to use as is? Other uses if modified? How would this product behave differently in another setting? Recycle the product to make something new?

Eliminate

Subtract? Smaller? Streamline? Simplify? Condense?

Reverse

Opposites? Turn backward? Upside-down? Mirror? Other layout? Other sequence? Change pattern? Change schedule?

For example, a fork and a spoon can be combined to create a unique eating utensil. Combining these units minimizes the number of utensils a person needs to use during a meal.

SCAMPER can also be applied to components of a project. For instance, the pockets of a backpack could be put to another use as a pencil or glasses container. The pockets could also be modified, moving them from the outside to the inside of the backpack or making them completely detachable.

Review Questions

1. Which of the following is NOT one of the four rules of brainstorming?

    A. Wild ideas are welcome.

    B. Quantity is more important than quality.

    C. Using others' ideas is discouraged.

    D. Criticism is not allowed.

2. Regarding the rule, "criticism is not allowed," who/what should you avoid criticizing?

    A. Your ideas

    B. Team member's ideas

    C. Your course instructor

    D. Both A and B

3. Which of the following is NOT a step included in the writing slip method?

    A. Individuals spend 15–20 minutes writing each idea on a notecard.

    B. Team members take turns explaining their ideas.

    C. Team members pass cards.

    D. Individuals hitchhike and add new cards to the stack.

4. During the card method _____.

    A. Teams iterate on only a few ideas

    B. Teams expand upon many ideas

    C. Team members remove cards from the deck if they do not think they will be successful ideas

    D. Team members select their favorite idea and pitch it to the rest of the team

5. Which two principles that govern idea generation were presented by James Webb Young? Select both answers.

   A. Brainstorming is the most effective process for idea generation.

   B. An idea is simply a new combination of old ideas.

   C. The ability to see relationships enhances one's capability to form new ideas.

   D. Quantity is more important than quality of ideas generated.

6. Which of the following is NOT one of the defining aspects of the SCAMPER method?

   A. Consists of divergent thinking strategies to expand the number of ideas generated.

   B. Eliminates detrimental effects of criticism in a group brainstorming session.

   C. Relaxes basic assumptions of an idea in order to open up discussion for more ideas.

   D. Results in an increased number of wild ideas.

Exercise #1

Task

Take 2 minutes to brainstorm possible uses for a flower pot (Figure 4.5). Try to generate as many ideas as possible.

Figure 4.5: Flower pot.

Select four ideas from your list and use the SCAMPER method to transform them into new ideas (Table 4.3). Be sure to identify which assumption was relaxed.

Original Idea	SCAMPER Word	New Idea	Assumption

Table 4.3: SCAMPER idea generation

Exercise #2

Introduction

An engineering design team is brainstorming possible uses for a paper grocery bag (Figure 4.6). After a few minutes, they came up with the following ideas.

Trash can   Luminaries

Mask   Book cover

Wrapping paper   To ripen fruit

Wallet   Storage containers

Task

List two or three assumptions that had could have been made about the paper grocery bag during the brainstorming process.

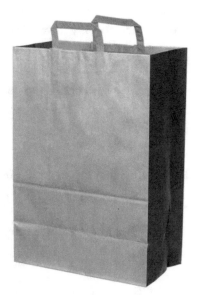

Figure 4.6: Paper grocery bag.

Eliminate these suspected prior assumptions from your mind. Using SCAMPER, generate two or three uses for a paper bag for each key word.

Substitute

Combine

Adapt

Modify

Put to other use

Eliminate

Reverse

Applying This Step to Your Project

Task: Generate Design Ideas: Round #1

Procure the necessary materials for brainstorming (e.g., index cards, pens, Post-it® notes, etc.) or use the following pages. Decide which idea generation method you will use and set a goal for the total number of ideas for this session (e.g., 30).

Begin a session that should last 30–45 minutes. When you start this brainstorming session, it should be quiet as you generate ideas independently. During sharing—either when discussing ideas or during formal hitchhiking—make sure to capture any new ideas by writing them down.

Task: Reflect on Round #1 Ideas

Reflect after your first session using the Brainstorming Tracker.

- Did you meet your goal for total number of ideas?

- Could your ideas be categorized as basic, repetitive, innovative, or far-fetched?

- Did you use hitchhiking to generate more ideas? Wild ideas?

- Did you consider all attributes, features, and functions of the design blocks (generated during decomposition) as launch points for idea generation?

Some ideas may be high level or abstract, but some ideas may likely be very practical and concrete. Prepare to address any deficiencies in your second session.

Task: Generate Design Ideas: Round #2

Meet for a second round of brainstorming that should last 30–45 minutes. Follow the steps above for an idea generation session and reflection. Address the deficiencies you discussed after your first session. Again, use the Brainstorming Tracker.

Task: Sort Generated Ideas

Sort the ideas into categories using the following pages. Categories could be identified based on a design block, function, overall structure, type of material, etc.

Brainstorming Tracker

<u>Round #1</u>

Time allocated:

Number of partial ideas:

Number of complete solution ideas:

<u>Reflection</u>

Did you meet your goal for total number of ideas?

Are your ideas varied and unique, or basic and repetitive?

Did you use hitchhiking to generate more ideas? Wild ideas?

Did you consider all attributes, features, and functions of the design blocks?

<u>Round #2</u>

Time allocated:

Number of partial ideas:

Number of complete solution ideas:

Solution Ideas

Use the boxes below to document brainstormed ideas. You can write or draw in the boxes, or you may place Post-it® notes in them. You may need more than 16 boxes.

Solution Ideas

Solution Ideas

Solution Ideas

Sort Generated Ideas

Sort your brainstormed ideas into categories based on design block, function, overall structure, etc. You may sort both partial and full solution ideas. You may need more than four categories.

Category:

Ideas:

Category:

Ideas:

Sort Generated Ideas

Category:

Ideas:

Category:

Ideas:

CHAPTER 5

Step 5: Evaluate Solutions

Evaluate solutions is the fifth step of the EDP (Figure 5.1). After brainstorming, teams will have numerous solution ideas that must be narrowed down to a couple viable designs. The strategy for narrowing ideas will depend on how many initial ideas were generated, as well as whether teams brainstormed full or partial solutions. Evaluating ideas with a systematic and quantitative method ensures that solutions with the best potential for satisfying design criteria are carried out in the prototyping phase.

Three substeps for evaluating solutions are discussed in this workbook. In Step 5A: Pugh Screening, the process of reducing many ideas down to a few using a Pugh screening matrix is explained. This tool for reducing ideas can be used for both full and partial ideas.

For projects where decomposition was used, there are often many component ideas that result from brainstorming. A morphological (or morph) chart can be used to combine these partial ideas into complete design solutions. Screening can occur both before and after using a morph chart. This morph chart is outlined in Step 5B: Morph Chart.

Once all solution ideas have been reduced to several viable, complete design solutions, a Pugh scoring matrix can be used. This matrix ranks the full solutions against established design criteria to determine the best solution for prototyping. Step 5C: Pugh Scoring describes this approach.

Evaluating solutions involves heavily discussing key functions and features of each design. As seen in Figure 5.1, this discussion can sometimes lead to teams revisiting the design criteria or repeating the brainstorming process.

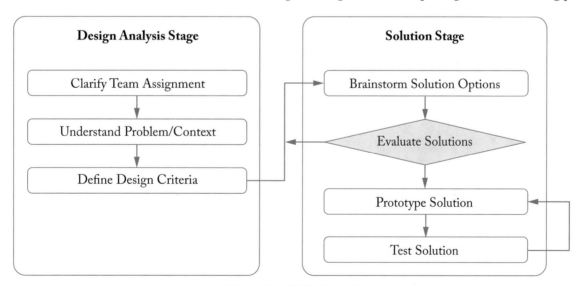

Figure 5.1: EDP: Step 5.

By the end of the section, you should be equipped to:

- Recognize different high-level strategies to sort through and evaluate ideas to select a candidate solution,

- Use a Pugh screening matrix to narrow solution ideas,

- Use a Pugh scoring matrix to rank solutions against established design criteria, and

- Use a morph chart to combine discrete, partial ideas into complete solutions.

STEP 5A: PUGH SCREENING

Playlist

Watch this video playlist for an overview about Pugh screening matrices. Learn how to systematically narrow brainstormed solutions for both full and partial ideas.

Videos	bit.ly/step5a-pughscreening

Video Notes

Define a Pugh screening matrix.

What are the three steps of engineering decision making?

When using a Pugh matrix, what does a plus (+) mean? What does a minus (-) mean?

How many solutions should you put in a Pugh matrix? Why?

What type of solution should be selected as the standard?

Content Summary

The purpose of Pugh screening matrices is to narrow down many solution ideas to a few ideas (Table 5.1). The steps to set up and run a Pugh screening matrix are as follows.

1. List the design criteria in the leftmost column.

2. List solution ideas across the top row.

3. Set a standard solution (has intermediate attributes compared with other solutions).

   • Place 0's down the column on the selected standard solution.

4. Evaluate the solutions—design criterion by design criterion.

   • (+) Improved relative to the standard

   • (0) Equivalent to the standard

   • (-) Worse than the standard

5. Add the scores for each solution.

6. Reflect on the results to combine or improve ideas.

7. Select concepts that move forward.

Table 5.1: Example Pugh screening matrix

Design Criteria	Solution A	Solution B	Solution C (standard)	Solution D	Solution E
DC 1	-	+	0	-	0
DC 2	+	+	0	-	0
DC 3	0	0	0	+	+
DC 4	+	+	0	-	0
Sum	1	3	0	-2	1
Rank	2nd	1st	3rd	4th	2nd

When running a Pugh screening matrix, make sure that you proceed carefully and objectively. If the resulting matrix has sums that fall mostly within 1 point of the standard, then the matrix should be redone with a different standard. Also, if the standard is the highest- or lowest-ranked solution, then the matrix should be redone with a different standard.

Review Questions

1. Decision matrices are evaluated using _____.

   A. A formula

   B. Preexisting design criteria

   C. Engineering software

   D. Iterative process diagrams

2. The numbers and quantitative data that inform decision matrices come from which of the following sources? Select all that apply.

   A. Research

   B. Expert opinions

   C. Experimental data

   D. Team member opinions

   E. Calculations

3. The main purpose of a Pugh screening matrix is to _____.

   A. Determine the best solution

   B. Increase the number of solution ideas

   C. Recombine solution ideas to form more complete ideas

   D. Reduce a large number of ideas to a few ideas

4. Which of the following is the standard score in a Pugh screening matrix?

   A. 0

   B. +

   C. –

   D. There is no "standard" value.

5. According to the sample screening matrix shown, which two solution ideas should continue forward to a Pugh scoring matrix?

	W	X	Y	Z
A	+	0	−	+
B	−	0	+	+
C	−	0	+	+

A. W and X

B. W and Y

C. Y and Z

D. X and Z

6. Chronologically organize the following steps for setting up a screening Pugh matrix.

A. Set a standard solution   1.

B. Reflect on the results; combine or improve ideas   2.

C. Add the scores   3.

D. List the design criteria in the leftmost column   4.

E. Evaluate the solutions design criterion by design criterion   5.

F. Select concepts that move forward   6.

G. List solution ideas across the top row   7.

7. How should you proceed through a Pugh Screening Matrix?

A. Moving down each column: considering one design solution against each design criterion.

B. Moving across each row: comparing each design solution to one design criterion.

C. Order does not matter as long as you assign each solution a -, 0, or +.

D. Order does not matter as long as you assign each solution a -1, 0, or 1.

8. During the reflection step, what are some things you might do? Select all that apply.

A. Check for compression of scores.

B. Determine which solutions move forward.

C. Check for an equal number of solutions above and below the standard.

D. Combine solutions.

Exercise #1

Introduction

An engineering design team developed the following design criteria for a giraffe hay feeder.

Safety   Durability
Feeding time   Asethetics
Food capacity

The team also collected the following information from their client interview and research.

- Metal surfaces are more likely to harm giraffes because of their sharp edges.

- Metal is known to be durable for 10 years. Plastics are durable for 5 years. Solid surfaces are more durable than mesh or other materials with many holes. Canvas is similar in durability to plastic.

- To blend in with an exhibit, the feeder should look like a tree. Color is important, and materials that are naturally brown, green or tan or can be painted one of those colors are more desirable. Metals that are painted retain their color much more readily than plastics that are painted.

- To extend feeding time, the surface area of the exposed hay must be less than 1–2 ft2.

Figure 5.2: Current hay feeder at the local zoo.

Task

Consider the following brainstormed ideas. Consider sketching them out.

Solution A: Holes in a plastic surface

Solution B: Holes in a metal surface

Solution C: Food delivered in small diameter tubes

Solution D: Wire mesh

Solution E: Metal grating (Figure 5.2)

Which of these ideas should serve as a standard? Why?

Complete Table 5.2 using the standard selected above.

Table 5.2: Pugh screening matrix for hay feeder					
	Solution A	**Solution B**	**Solution C**	**Solution D**	**Solution E**
Safety					
Feeding time					
Food capacity					
Durability					
Aesthetics					
Sum					
Rank					

Which solutions should move on for further consideration? Which should be eliminated?

Exercise #2

Introduction

An engineering team was tasked with designing an eating utensil for children with musculoskeletal disorders. After brainstorming, the team decided to narrow solutions by screening them against five design criteria. The Pugh screening matrix (Table 5.3) lists the design criteria in order of importance.

Task

Complete the Pugh screening matrix in Table 5.3.

Table 5.3: Pugh screening matrix for eating utensil															
	A	B	C	D	E	F	G	H	I	J	K	L	M	N	O
DC 1	+	+	0	–	+	0	–	+	0	0	+	–	+	+	0
DC 2	0	0	0	0	0	0	0	0	0	0	0	0	0	0	0
DC 3	+	0	–	+	+	0	–	–	+	+	0	+	0	+	–
DC 4	–	+	–	0	+	0	–	–	0	–	+	0	+	0	–
DC 5	+	–	0	+	+	0	–	0	–	0	+	–	0	–	+
Sum															

Which solution is the standard?

Which solutions should be eliminated before moving to another decision matrix?

What could the team improve about this matrix?

Applying This Step to Your Project

Sort all brainstormed solutions into two categories: full solutions and partial solutions. Component/partial solutions may need to be screened independently before being combined to create full solutions in a morph chart. The screening process is iterative and unique for each team, depending on how many solutions have been brainstormed. The flowchart in Figure 5.3 shows different paths a team may take when evaluating solutions.

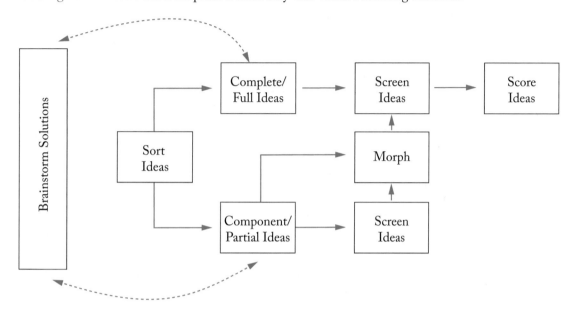

Figure 5.3: Flowchart for engineering decision making.

Task: Screen Complete/Full Ideas Using a Screening Pugh Matrix

The goal of this task is to reduce many complete/full ideas down to 5–10 ideas using a series of screening matrices.

- Screen your ideas based on reality and common sense. Throw out ideas that are ridiculous or impractical.

- Complete a Pugh screening matrix for a first set of ideas. You should aim to eliminate 50% or more of your ideas.

- Repeat Pugh screening matrices until all ideas have been vetted. There should only be about 5–10 ideas that remain.

Task: Screen Component/Partial Ideas Using a Screening Pugh Matrix

The goal of this task is to reduce many component/partial ideas by screening to prepare for a morph chart.

- Confirm that your component/partial ideas are sorted into design blocks that can be grouped by like parts, features, or components. Examples of design blocks are "handles" or "lifting mechanisms."

- Screen ideas in each design block based on reality and common sense. Eliminate ideas that are ridiculous or impractical.

- Complete a Pugh screening matrix for the first design block of component/partial ideas. Eliminate 50% or more of your ideas.

- Repeat Pugh screening matrices until each design block has been vetted and about three to four ideas remain in each block.

Use the following blank Pugh screening matrices to evaluate your brainstormed solutions.

Table 5.4: Blank screening matrix

Design Criteria								
DC 1:								
DC 2:								
DC 3:								
DC 4:								
DC 5:								
DC 6:								
Sum								
Rank								

Table 5.4: Blank screening matrix

Design Criteria								Sum	Rank
DC 1:									
DC 2:									
DC 3:									
DC 4:									
DC 5:									
DC 6:									

STEP 5B: MORPH CHART

Playlist

Watch this video playlist for an overview about morph charts. Learn how to combine component ideas and partially brainstormed solutions into full designs.

Videos	bit.ly/step5b-morphcharts

Video Notes

Define a morphological chart.

How are solutions categorized before being placed in a morph chart?

How many ideas could be selected from each design block?

When should a team stop generating solutions using a morph chart?

How are morph charts different from Pugh screening and scoring decision matrices?

Content Summary

Morph charts are visual tools used to combine partially formed design ideas into complete solution ideas. Teams can set up and use morph charts by following these steps.

1. Organize ideas into design blocks.

2. Write the name of each design block on the leftmost column of the chart.

3. For each design block, fill in each row with the brainstormed ideas.

4. Create a full solution by selecting zero, one, or two ideas from each of the design blocks and combine them into a novel solution idea.

5. Repeat until you generate 8–15 complete solution ideas that cover the design space and are unique from one another.

An example Morph chart is shown in Table 5.6.

Table 5.6: Morph chart for water bottle design			
Design Block	**Option 1**	**Option 2**	**Option 3**	**Option 4**
Mouth piece	Twist top	Spigot	Rubber nipple	Pull top
Container	Plastic	Disposable	Metal	
Handle	Bottle modified	Backpack	Loop	None
D:H ratio	<1:2	1:2–1:3	>1:3	
Shape	Ergonomic	Pouch	Constant diameter	Ribbed

Review Questions

1. Which of the following best describes the purpose of morph charts?

   A. Visual tool to combine partially formed design ideas into full solutions

   B. Ranking system to score brainstormed ideas

   C. Chart to prioritize design objectives

   D. Visual tool to clarify the major components of the design challenge

2. The design block categories should be listed along _____.

   A. A column

   B. Each row

   C. The first column

   D. The top row

3. Consider a design block with two compatible ideas. How many ideas could be selected from that design block?

   A. One

   B. One or two

   C. None or one

   D. None, one, or two

4. Morph charts are especially effective for teams that _____.

   A. Disagree on which proposed design is best

   B. Have many incomplete design ideas

   C. Struggle with ranking their brainstormed ideas

   D. Don't have enough incomplete design ideas

5. A team should continue using a morph chart until _____.

   A. There are enough assembled design solutions covering a broad range of the design space

   B. They have reached 20–25 complete solutions

   C. A complete and feasible solution is created

   D. The client is satisfied with one of the design solution ideas

6. Use the morph chart for the design of a globe to determine which answer choice cannot be a solution.

Design Block	Option 1	Option 2	Option 3
Stand	Platform	Hanging	Tripod
Features	Color coding	Tactile raised surface	City labels
Orientation	North pole up	South pole up	Tilted

   A. Platform, south pole up, color coding

   B. North pole up, south pole up, tactile raised surfaces

   C. Hanging, color coding, city labels

   D. Tripod, color coding, north pole up

Exercise #1

Introduction

A design team is tasked with designing a new dual-purpose backpack. The team came up with the following brain-stormed ideas for the design.

String shoulder straps	Leather
Fabric shoulder straps	Canvas with leather bottom
Padded fabric shoulder straps	Various synthetic plastics
Biodegradable	Metal zippers
Computer holder	Plastic zippers
Pencil holder	2 compartments
Small compartments	3 compartments
Water bottle attachment	4 compartments
Canvas	5 compartments

Task

Organize the above brainstormed ideas into discrete design blocks. Use these blocks to set up a morph chart in Table 5.7.

Table 5.7: Morph chart for backpack project

Design Block	Option 1	Option 2	Option 3	Option 4	Option 5

Exercise #2

Introduction

An engineering team is tasked with creating a flexible device for positioning phototherapy lights at a hospital. The team sorted their brainstormed solutions into discrete design blocks and created the morph chart in Table 5.8.

Table 5.8: Morph chart for phototherapy lights project				
Design Block	**Option 1**	**Option 2**	**Option 3**	**Option 4**
Attachment Mechanism	Clip	String	Magnet	Permanent
Adjustment Mechanism	Slider	Crank	Pulley	Telescoping tubing
Material	Wood	Sheet metal	Plastic	
Lights	Row of LEDs	Incandescent bulb	Cluster of incandescent bulbs	

Task

Create eight to ten diverse design solutions from the morph chart.

Are there any solutions that are not feasible? Why?

Applying This Step to Your Project

Task: Complete Morphological Chart(s)

The goal of this task is to create full design solution ideas by combining component/partial solutions from various design blocks.

- Evaluate whether "full" solutions can be constructed from the component/partial ideas in design blocks using a morph chart. If this is not possible, you may need to spend additional time brainstorming. On the other hand, it may make sense for these partial solution ideas to be add-on features.

- If you have more than six ideas in a design block, use the steps outlined in Step 5A: Pugh Screening to reduce the ideas in that design block.

- Construct and complete a morph chart to generate complete/full solution ideas.

Task: Screen Ideas

Given the large number of new ideas, your team may need to vet all newly developed full solution ideas by following the steps outlined in Step 5A: Pugh Screening.

Use Table 5.9 to construct a Morph chart.

Table 5.9: Blank morph chart

Design Block	Option 1	Option 2	Option 3	Option 4	Option 5	Option 6

STEP 5C: PUGH SCORING

Playlist

Watch this video playlist for an overview about the Pugh scoring matrix. Learn how to quantitatively decide on a final solution for moving forward in the prototyping process.

Videos	bit.ly/step5c-pughscoring

Video Notes

Define a Pugh scoring matrix.

How is the Pugh scoring matrix similar to the Pugh screening matrix? How are these matrices different?

How should the weights for design criteria be determined?

List three questions you should ask yourself when reflecting on the scoring process.

Content Summary

The purpose of Pugh scoring matrices is to narrow down potential solutions to one or two ideas (Table 5.10). The steps to set up and run a Pugh scoring matrix are as follows.

1. List the design criteria in the leftmost column.

   • Add weights in the column immediately to the right; weights should be >5%.

2. List solution ideas across the top row.

3. Set a standard.

   • Each design criterion has its own standard solution.

   • Standards are given a "3."

4. Evaluate the solutions—design criterion by design criterion.

   • For each design criterion, compare the solution to the standard and give the solution a number between 1 and 5.

5. Add the scores for each solution.

   • Multiply the rating by the weight, then sum the column.

6. Reflect on the results to combine or improve ideas.

   • Scoring matrices are only effective when the sums have differences of >0.2.

7. Select concept(s) as the final solution idea.

Table 5.10: Example of a Pugh scoring matrix

Design Critera	Weight	Solution A/E		Solution C		Solution D		Solution G		Solution J	
		Rating	Weight Score	Rating	Weight Score	Rating	Weight Score	Rating	Weight Score	Rating	Weight Score
Time	30%	2	0.6	3	0.9	5	1.5	3	0.9	5	1.5
Transport	20%	2	0.4	3	0.6	4	0.8	1	0.2	5	1
Ease of Use	15%	2	0.3	3	0.45	2	0.3	1	0.15	3	0.45
Light-weight	15%	3	0.45	1	0.15	3	0.45	5	0.75	1	0.15
Durable	10%	2	0.2	5	0.5	2	0.2	3	0.3	3	0.3
Cost	10%	1	0.1	4	0.4	3	0.3	4	0.4	3	0.3
Total Score	100%	2.05		3.00		3.55		2.70		3.70	
Rank		5		3		2		4		1	
Continue		No		No		No		No		Yes	

Using Proxies for Decision Matrices

Sometimes abstract ideas that have not been made into solutions are difficult to evaluate against design criteria. For instance, durability is a design criterion that is important for many engineering projects. However, aspects of a project that impact durability such as material selection, safety factors, and manufacturing processes are not decided until the prototyping stage. However, it is still important to find a way to objectively define durability for solution concepts.

Proxies can be used to define design criteria that are not concrete. A proxy is merely a descriptor that stands in place of the design criterion. A proxy for durability could be the number of moving parts, number of failure points, or number of components that must be machined. By transforming the criterion of durability into a discrete proxy, abstract solutions can be described before they are even created. Using proxies allows the decision-making process to remain objective and quantitative, reducing opinions and guesswork.

Some further examples of design proxies, as well as how those proxies would be used in a scoring matrix, are listed in Table 5.11.

Table 5.11: Example proxies

Design Criterion	Proxy	Scoring Matrix Ranking
Durability	Number of parts that required advanced machining	1: 6+ parts 2: 4–5 parts 3: 2–3 parts 4: 1 part 5: 0 parts
Ease of Use	Number of steps for assembly	1: 10+ steps 2: 8–9 steps 3: 5–7 steps 4: 2–4 steps 5: 1 step

Note: It is okay for the 1–5 ranking to be different from the target value for the design criterion. This scale ranking will not be used to evaluate the final solution. It is only used for the scoring matrix.

Review Questions

1. Which functions of the Pugh scoring matrix are the same as the Pugh screening matrix? Select all that apply.

   A. Rank solutions using weighted design criteria.

   B. Rate concepts against design criteria.

   C. Combine/improve solutions.

   D. Select solutions for further evaluation.

2. The following intervals are used to define the ratings for design criteria. Which numbers should be placed in the definition of the "3" rating?

1	>$40
2	$30–40
3	
4	$10–20
5	<$10

   A. $0–10

   B. <$40

   C. $35

   D. $20–30

3. Screening matrices use a _____scale, while scoring matrices use a _____scale.

   A. Two-point; five-point

   B. Three-point; five-point

   C. Two-point; three-point

   D. Three-point; ten-point

4. Organize the following steps into the correct order.

   A. Evaluate the solutions.   1.

   B. Set a standard.   2.

   C. Reflect.   3.

   D. List design criteria and weighted percentages.   4.

   E. List solution ideas.   5.

   F. Select concepts that move forward.   6.

   G. Add the scores.   7.

5. Which of the following statements about scoring Pugh matrices are true? Select all that apply.

   A. Design criteria are assigned weights.

   B. Weightings are typically in multiples of five.

   C. Design criteria with weights less than 5% should be included.

   D. Weightings are percentage values.

6. Which of the solution ideas is the "standard" for design criterion A in the scoring matrix shown?

Design Criteria	Weight (%)	Solutions			
		W	X	Y	Z
A	30	1	5	3	2
B	35	3	1	5	4
C	35	2	2	5	3

   A. W

   B. X

   C. Y

   D. Z

7. What is the total score for solution W in the scoring matrix shown?

Design Criteria	Weight (%)	Solutions			
		W	X	Y	Z
A	30	1	5	3	2
B	35	3	1	5	4
C	35	2	2	5	3

   A. 4.3

   B. 3

   C. 2.7

   D. 2

8. What is wrong with the row for design criterion B in the scoring matrix shown?

Design Criteria	Weight (%)	Solutions			
		W	X	Y	Z
A	55	3	3	3	3
B	40	2	3	3	2
C	5	1	2	4	5

A. The design criterion is not weighted enough.

B. The design criterion is weighted too little.

C. There is no standard solution.

D. There is not a large enough range in ratings.

Exercise #1

Introduction

An engineering design team developed the following design criteria for a giraffe hayfeeder (Figure 5.4).

Feeding time   Durability
Food capacity   Aesthetics

Task

Create proxies for the design criteria. Be sure to indicate descriptions for the levels 1–5 for the Pugh scoring matrix.

Figure 5.4: Giraffe hayfeeder.

Feeding time:

Food capacity:

Durability:

Aesthetics:

Exercise #2

Introduction

A design team has generated Table 5.12 to evaluate five final solutions.

Design Criteria	Weight	Solutions				
		V	W	X	Y	Z
A	0.30	2	3	1	5	4
B	0.30	1	4	3	2	2
C	0.30	1	2	1	2	3
D	0.05	3	2	4	4	2
E	0.05	1	5	2	1	4
Score						

Table 5.12: Pugh scoring matrix

Task

Complete the matrix by weighting each solution with the design criteria and summing the scores.

Is there a standard for every design criterion? If not, which design criterion does not have a standard?

Consider the design criteria A, B, and C. Is the whole range (1–5) used for each design criterion? How could this affect the selected solution and utility of the Pugh scoring matrix?

Should the team proceed with this matrix? If not, explain what could be improved in the matrix.

Applying This Step to Your Project

Task: Select a Final Solution Using a Scoring Pugh Matrix

The goal of this task is to reduce several complete/full solutions to a final solution.

- Complete one or two Pugh scoring matrices to reduce the best five to ten ideas (from various Pugh screening matrices) to one or two ideas.

- Write the final Pugh scoring matrices in Table 5.13 and Table 5.14 for reference throughout the rest of the project.

Table 5.13: Blank Pugh scoring matrix

Design Criteria	Weight	Solution: Rating (1–5)	Solution: Score	Solution: Rating (1–5)	Solution: Score	Solution: Rating (1–5)	Solution: Score	Solution: Rating (1–5)	Solution: Score	Solution: Rating (1–5)	Solution: Score
Total Score	100%										
Rank											

Table 5.14: Blank Pugh scoring matrix

Design Criteria	Weight	Solution:		Solution:		Solution:		Solution:	
		Rating (1–5)	Score	Rating (1–5)	Score	Rating (1–5)	Score	Rating (1–5)	Score
Total Score	100%								
Rank									

CHAPTER 6

Step 6: Prototype Solution

Prototyping is the sixth step of the EDP (Figure 6.1). Prototyping is the process of creating a solution to a design challenge. The goals of prototyping are to test an idea or concept, explore brainstormed ideas, compare potential design solutions, communicate design ideas, and/or evaluate a feature. For example, suppose a team is tasked with creating a hand-washing station for use in a low-income community. One prototype could be non-functional and show the overall shape of the station, while another prototype could demonstrate the effective flow of water. The iterative nature of prototyping and exploring aspects of the design is key to developing a successful solution. The process of evaluating a prototype's success is described in the following step.

This workbook contains three related substeps for prototyping. Step 6A: Safety explains guidelines for creating a safe working environment. Step 6B: Initial Prototyping provides information on how to start prototyping, including how to evaluate preliminary prototypes. The final substep, Step 6C: Refined Prototyping, explains how to refine successive prototypes. It lists advice for increasing fidelity of prototypes and improving design functions.

Depending on the project, it could be helpful to evaluate the prototype through rigorous testing (outlined in Step 7: Test Solution) before proceeding to Step 6C: Refined Prototyping. Prototyping and testing are iterative processes that can occur sequentially or simultaneously. This is seen in Figure 6.1 by the feedback arrow between Step 6: Prototype Solution and Step 7: Test Solution.

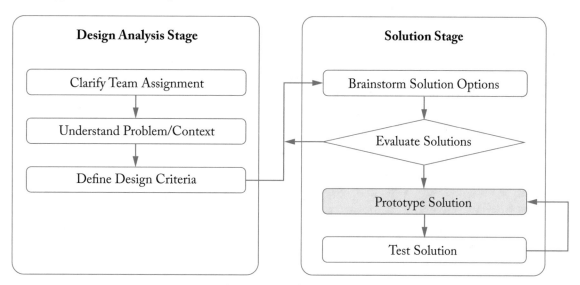

Figure 6.1: EDP: Step 6.

By the end of the section, you should be equipped to:

- Practice safety guidelines for working with prototyping equipment,

- Identify the goals of the four types of prototypes (e.g., representational, functional),

- Build iterative, meaningful, and successive prototypes,

- Differentiate between low-, medium-, and high-fidelity prototypes, and

- Identify areas for increasing fidelity in a prototype.

STEP 6A: SAFETY

Playlist

Watch this video playlist for an overview about safety when prototyping. Learn about the culture of safety and how it can prevent prototyping accidents.

Videos	bit.ly/step6a-safety

Video Notes

List examples of small-scale, medium-scale, and large-scale accidents.

When are accidents most likely to occur?

What is intervention? When should you intervene?

What is a near miss? List some examples of near misses with prototyping.

Content Summary

A culture of safety is vital for preventing accidents during the prototyping process. Although most accidents are small-scale and can be remedied with common first aid supplies, larger-scale accidents that require professional medical attention can occur. Accidents are most likely to occur when designers are:

- Prototyping too fast,

- Taking a shortcut,

- Working while tired,

- Not focusing on the task at hand,

- Trying something new,

- Breaking the rules, and

- Using a tool incorrectly.

A strong safety culture is created when designers always consider safety first before performing any action, intervene when others are practicing improper safety procedures, and reflect on accidents that do occur. This reflection is equally important for near misses, instances where an accident almost happened but did not.

Proper protective equipment (PPE) must also be worn to ensure safety during prototyping. PPE includes gloves, safety glasses, closed-toe shoes, and pants. Long hair and dangling jewelry (such as necklaces) should be removed or fastened in such a way that they cannot interfere with using tools.

Practice a culture of safety by always working with a partner, never working when tired or distracted, asking questions, and wearing PPE.

Additional safety measures may be required based on local rules. Please be observant of all course-based and facility-based rules for remaining safe while working.

Review Questions

1. When can accidents occur? When you are _____. Select all that apply.

   A. Tired

   B. Going too fast

   C. Using a tool incorrectly

   D. Breaking the rules

   E. Trying something new

   F. Complacent about an activity, due to boredom or repetition

   G. All of the above

2. In order to develop skills in intervention, you need to be able to see ____.

   A. Yourself

   B. Others

   C. Neither yourself or others

   D. Both yourself and others

3. A near miss might include:

   A. A cut on your finger

   B. A burn on your arm

   C. A tool that slips but doesn't cut your hand

   D. Using the first aid kit

4. PPE does NOT include ____.

   A. Gloves

   B. Sandals

   C. Long pants

   D. Safety glasses

5. You should wear safety glasses when working with which of the following tools and materials?

   A. Liquids

   B. Electronics

   C. Hand tools

   D. All of the above

6. Which of the following are aspects of a safety culture? Select all that apply.

   A. Intervention

   B. Demonstration

   C. Reflection

   D. Proper personal protective equipment

7. This student was caught trying to use the drill press (Figure 6.2). Select ALL safety procedures she is NOT doing.

Figure 6.2: Unsafe student.

A. Wearing proper shoes

B. Wearing safety glasses properly

C. Wearing an appropriate hairstyle

D. Keeping loose clothing away from the machine

Exercise #1

Task

For each of the following scenarios, identify the safety concern. How would you intervene in this situation? How could this concern be prevented in the future?

1. One of your team members refuses to wear closed-toe shoes when prototyping. He complains that they make him uncomfortable, so he only wears sandals.

2. Your teammate prototypes every night after 10 pm. She explains she has a packed schedule between classes and a part-time job, so she can only prototype late in the evening.

3. While attempting to fasten two pieces of wood together, your team member loses his grip and narrowly avoids pounding his finger with a hammer. He continues to prototype as if nothing happened.

Applying This Step to Your Project

Task: Determine Safe Working Conditions

The goal of this task is to create a Safety Plan to define a culture of safety in your working environment. By outlining guidelines for prototyping, you can prevent design accidents before they occur. The Safety Plan document should address the following questions.

- Who dictates what rules must be followed for safe prototyping? Where are these rules posted?

- What PPE should be used for this project? List the specific tools and prototyping processes to be used. What specific protections are required?

- Where is the first aid kit located? Where is the fire extinguisher?

- What steps should be taken after near miss events?

- How should prototypes and tools be stored?

- What standards should there be for cleanliness? How often should the prototyping area be cleaned?

STEP 6B: INITIAL PROTOTYPING

Playlist

Watch this video playlist for an overview about the classification of prototypes. Learn the purposes of different prototypes and how to begin designing your solution.

Videos	bit.ly/step06b-initialprototyping

Video Notes

Define the following key terms.

- Low-fidelity prototype:

- Representational prototype:

- Functional prototype:

- Ergonomic prototype:

- Proof-of-concept prototype:

- Medium-fidelity prototype:

- High-fidelity prototype:

What are the purposes of prototyping?

Which kinds of prototype are made to-scale? Why is this scaling important?

How much time should be spent completing initial prototypes?

Content Summary

Prototyping is the iterative process of creating a solution to a design challenge. The goals of prototyping are to test an idea or concept, explore brainstormed ideas, compare potential design solutions, communicate design ideas, and/ or evaluate a feature.

There are four different types of prototypes with distinct goals (Figure 6.3).

- Representational prototype: a non-working model that symbolizes design blocks and features

- Functional prototype: a working model whose features match the intended utility or action

- Ergonomic prototype: a model that addresses how people interact with the device

- Proof-of-concept: a model that explores uncertain ideas, focusing on the question "does it work?"

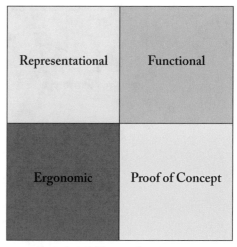

Figure 6.3: Types of prototypes.

Throughout the prototyping process, your team should iterate and evolve on both the prototype's components and the integrated design (Figure 6.4).

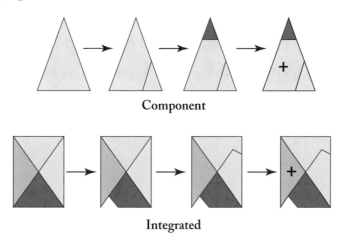

Figure 6.4: Potential prototype evolution.

As the prototyping stage progresses, prototypes increase in fidelity or resolution. A low-fidelity prototype is built quickly to test ideas. Low-fidelity prototypes rarely resemble the final product. A medium-fidelity prototype may have functional components, satisfy some design criteria, and/or have some integrated design blocks. These prototypes begin to reflect the manufactured product.

Finally, a high-fidelity prototype has functional components, satisfies most design criteria, and has integrated design blocks. High-fidelity prototypes may not be produced with the same method or with the same materials as a manufactured product. Figure 6.5 shows a general timeline for the prototyping process.

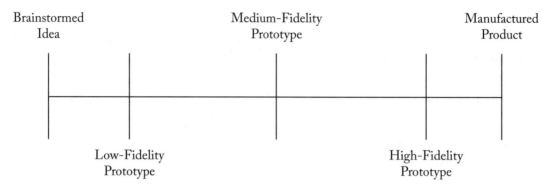

Figure 6.5: Progression of fidelity of prototypes.

Review Questions

1. Low-fidelity prototypes differ from high-fidelity prototypes in what ways? Select all that apply.

    A. They are made from low-cost rather than high-cost materials.

    B. They contain artisan-based defects while high-fidelity prototypes do not.

    C. They are built to test all initial desired functions while high-fidelity prototypes are usually built only to test individual functions.

    D. They are built rapidly while high-fidelity prototypes are built with more time and precision.

    E. They are often not built to scale while high-fidelity prototypes are often built to scale.

2. With each successive degree of fidelity ____.

    A. The number of working functions increases

    B. Artisan-based defects become less and less noticeable

    C. The prototype more closely resembles the desired product

    D. All of the above

3. Which of the following are purposes for prototyping? Select all that apply.

    A. Communicate design ideas

    B. Test a feature

    C. Test an idea or concept

    D. Explore brainstormed ideas

    E. Compare potential design solutions.

4. Prototypes can evolve in their ____.

   A. Complexity

   B. Functionality

   C. Fidelity

   D. All of the above

5. Which prototype is used to address how people interact with the device?

   A. Representational

   B. Ergonomic

   C. Proof-of-concept

   D. Functional

6. Initial prototypes should be constructed in ____.

   A. <15 min

   B. <30 min

   C. <1 hr

   D. <1 day

7. When starting to build a prototype, your team might start with ____.

   A. Paper and tape

   B. Cardboard and PVC

   C. Flow chart

   D. All of the above

8. Immediately after finishing a prototype, what should you do?

   A. Redesign every component.

   B. Order new materials online.

   C. Evaluate the prototype.

   D. None of the above.

Exercise #1

Introduction

Your team has been tasked with developing a new spill-proof coffee mug. Your team has used Pugh screening and scoring matrices to narrow down the ideas to one conceptual design. The characteristics of this design are listed below.

Rubber grip   Cylindrical shape

Liquid leaves mug when a button is pushed   Task

Double-walled

Task

Using this information, take 5 minutes to sketch your understanding of the coffee mug design below.

If possible, partner with a neighbor and present your sketches to each other. Then, evaluate the sketches. Which characteristics lacked clarity? What characteristics should be added to the description of the solution to make it more complete? If you are working with a partner, compare your sketches. How are they similar and different?

Exercise #2

Introduction

Your team is tasked with creating a backpack for hiking and everyday wear that is lightweight and weather resistant. The backpack should also be comfortable and aesthetically pleasing. More information on this project is given in Step 2: Understand the Problem and Context.

Task

Write a description of a prototype for each of the following categories. Identify at least two low-fidelity materials that could be used to build these prototypes.

Representational prototype:

Functional prototype:

Ergonomic prototype:

Proof-of-concept prototype:

Applying This Step to Your Project

Task: Create a Tools and Materials Inventory

It is helpful to understand what tools, machines, and processes are available for building prototypes. Inventory your workspace. Be sure to answer the following questions.

- What tools, materials, machines, software, etc. are available for this project?

- What tools, materials, machines, software, etc. would I like to acquire for prototyping?

- What additional skills do I need to learn before using a specific tool?

- Based on this inventory, do I need to update my Safety Plan?

Note that you may need more rows than are provided in Table 6.1.

Table 6.1: Workspace inventory

Tool/ Material	Location	Quantity

Task: Draw, Describe, and Discuss the Design

The purpose of an initial discussion is to come to consensus on the solution, represented through a series of sketches. As a team, first discuss how you envision your final design in detail. Use your design blocks to guide the discussion, but do not forget about how you will integrate them. When starting this step, focus on the "fuzzy" (less well-specified) or "hardest" (technically challenging) parts first. The design sketches should:

- Illustrate various pieces, parts, flows, etc.,

- Be hand drawn (avoid CAD sketches at this step as they will slow your team down unnecessarily),

- Include separate sketches for intricate aspects of the solution and/or important design blocks,

- Be drawn with consideration to scale and include approximate dimensions,

- Include different views (e.g., front perspective, side perspective, inside, outside), and

- Focus on flowcharts and/or pseudocode for coding or electrical design challenges.

Depending on the clarity of your design, your team may use different methods to draw and describe the proposed design. If there is a good consensus about what the design will look like, one person can draw while the other team members ask questions to clarify and refine. If there is little to no consensus, each team member should draw what they believe the design will look like.

With this discussion, it is important to be frank and identify the parts of the design you don't currently understand. The goal is that any person from the team should be able to clearly describe the solution with the completed drawings.

Design Sketches

Design Sketches

Task: Start to Prototype

The team should start by constructing a prototype with low-fidelity materials. If your team feels stuck, here is a range of strategies that could help you begin.

- Look at the low-fidelity materials available for prototyping.

- Borrow ideas from an existing device. For example, if you need a flap that opens and closes, look at existing hinges. If you need a thin bundle that expands radially, look to see how an umbrella works.

- Select some materials and fasteners for a section or design block of your prototype. Start prototyping this section or block.

- Discuss and decide on the scale of your prototype. Should you build big or small?

- Storyboard a user's interaction with your device. How and in what order does a user work with your device?

- Develop a flowchart of your solution. This strategy is particularly helpful for prototypes with electronics or flowing liquids.

- Physically manipulate relevant objects. For example, if your team is making an object that uses a bar of soap, get a bar of soap to touch and feel.

- Think about similar solutions to your problem. How are they constructed? What materials are they made of? How are those devices put together?

Remember that you should:

1. Bias toward action. Don't be afraid to start prototyping once your team has a shared understanding of the design.

2. Take <30 minutes to construct your first low-fidelity prototype.

3. Talk about what parts you think will be the most difficult to build or might fail first.

4. Build each prototype with a clear goal in mind.

Task: Complete Your First Prototype

Celebrate the completion of your first prototype. Take several pictures from different perspectives. Complete the first column of Table 6.2: Prototyping Log. Record the purpose, strengths, and weaknesses of this first prototype. Identify some goals for your second prototype.

Task: Complete Additional Prototypes

As you refine design concepts, you will iterate through many design solutions. It is important to document the purpose of each prototype as well as the outcome. Did the solution prove a fundamental design concept? Did the solution provide insight into a possible safety concern? Build your second, third, and fourth prototypes. Take several pictures of each and complete the prototype logs on Tables 6.2–6.3.

Table 6.2: Prototyping log

Prototype #	Prototype #
Circle the category of prototype Representational Functional Ergonomic Proof-of-concept	Circle the category of prototype Representational Functional Ergonomic Proof-of-concept
Sketch/description of prototype (Include link to source image or video)	Sketch/description of prototype (Include link to source image or video)
Strengths of prototype	Strengths of prototype
Weaknesses of prototype	Weaknesses of prototype
Goals for next prototype	Goals for next prototype

Table 6.3: Prototyping log

Prototype #	Prototype #
Circle the category of prototype Representational Functional Ergonomic Proof-of-concept	Circle the category of prototype Representational Functional Ergonomic Proof-of-concept
Sketch/description of prototype (Include link to source image or video)	Sketch/description of prototype (Include link to source image or video)
Strengths of prototype	Strengths of prototype
Weaknesses of prototype	Weaknesses of prototype
Goals for next prototype	Goals for next prototype

STEP 6C: REFINED PROTOTYPING

Playlist

Watch this video playlist for an overview about medium-fidelity prototypes. Learn how to refine prototypes and increase fidelity of materials.

Videos	bit.ly/step06c-refinedprototyping

Video Notes

<u>Answer the following questions regarding Team Safe Soap.</u>

What are four examples of medium-fidelity fasteners?

What criterion was used to decide between two design solutions? Why was this important?

List one weakness of a prototype and how it informed the next iteration.

<u>Answer the following questions regarding Team IncuBaby.</u>

What material was used for the initial prototype? Why was this selected?

List three "pro tips" outlined in the video.

Content Summary

The goal of iterative prototyping is to improve and evolve the key functions of the design solution. Iterative prototyping consists of active evaluation and critique with incremental improvements. As you enhance your prototype, your solution should increase in fidelity and quality of materials. For instance, cardboard may be replaced by wood or a circuit board replaced by a PCB. The quality of attachments in a prototype should also increase (e.g., masking tape replaced by finger joints; glue replaced by hinges).

While the time spent prototyping increases over the course of the project, the number of physical prototypes actually decreases. You should spend more time on each medium- to high-fidelity prototype, improving aspects and refining features. The improvements needed are often informed by rigorous testing of the prototype against design criteria. This is outlined in Step 7: Test Solution. As you work on refining your solution, you should consider the following.

- Scale: When is it appropriate to build at a relevant scale (or much larger or much smaller)?

- Design blocks: Should the team focus on one design block at a time, or can the team work on several design blocks simultaneously?

- Core functions: Which core functions are most important, and how can meeting those functions be prioritized as your team prototypes?

- Making vs. buying: How can you speed up prototyping by finding or purchasing some built objects or materials? What must be made? What can be modified from an existing product?

- Design Criteria: Which design criteria are most important to meet, and how can your prototyping strategies be prioritized to meet those criteria (see Step 7: Test Solution)?

Some additional tips during this phase of prototyping include:

- Let function drive the design,

- Use a material that is easy to work with,

- Don't reinvent existing products; incorporate them into your design,

- Critically evaluate each prototype,

- Iterate to alter one feature at a time,

- Get client and advisor feedback early,

- Expect and embrace failure,

- On large-scale projects, build small to iterate fast, and

- Label your prototypes.

Sometimes a team will slow down or stop prototyping and no progress on the design solution will be made. This can occur before the solution has been completely refined, and often before all the design criteria are fully satisfied. Table 6.4 lists several common reasons teams may stop prototyping and suggestions for getting back on track to finish the design project.

Table 6.4: Solutions for common prototyping challenges

Reason for Stopping	Suggestions
Lack of tools	• Visit a local makerspace • Talk to a carpenter, electrician, plumber, or engineer in your community • Visit a hardware store
Lack of funding	• Talk to a teacher • Look for local organization funds • Use tools and materials that surround you
No assistance available	• Look for a local professional engineering society (e.g., ASME) • Look for guilds or machine shops • Find a technical advisor
Uncertain how to assemble or build project	• Visit a hardware store • Find a technical advisor • Commit to trying and failing anyway
Client changes scope of problem or what they wanted	• Talk to client to redefine design criteria and evaluate current prototype • Document for another team to take over the project
Out of time	• Document for another team to take over the project • Deliver to client with future plans outlined

Exercise #1

Introduction

Integration is an important step in the prototyping process, and it is often the most difficult for teams to complete. Determine how to integrate this team's component parts into a complete solution. This may require a complex joining operation of two or more materials. Or, it could require the use of a custom designed or built device.

Task

For each of the design challenges in Table 6.5, describe how the component parts could be integrated. There may be more than one correct integration method.

Table 6.5: Integration of component parts		
Design Challenge	**Component Parts**	**Method for Integration**
Create an insulated coffee mug	• Cylindrical, steel mug • Ergonomic plastic sleeve	
Create a backpack that can be used for everyday activities and hiking	• Exterior mesh pocket • Nylon backpack straps	
Create a hay feeder for giraffes at the local zoo	• Rectangular metal frame • Fabric straps for hanging feeder from stand	
Create a hand-washing station for a low-income community	• Rectangular wooden frame • Plastic tubing that moves water from reservoir	

Applying This Step to Your Project

Task: Self-Evaluate Working Prototypes

Each prototype is created for a specific purpose, whether to demonstrate a physical concept or to represent a desired feature. It is critical to evaluate each prototype to understand the strengths and weaknesses. Any limitations or design errors will inform how future prototypes should be constructed. When refining prototypes, be sure to answer the following questions.

- What is the intended purpose of this prototype?

- What function is the prototype demonstrating?

- What weaknesses can be improved or refined?

- What are the next steps?

Under next steps, indicate what new materials or advanced prototyping processes could be used in future iterations. Complete the evaluation logs on Tables 6.6–6.7 for each working prototype.

Task: Evaluate Prototype by an External Reviewer

Although self-evaluation is important, prototypes can benefit from review by a person who is not a part of the project. External critiques can help teams identify new areas for improvements and be especially helpful when teams are fixated on a design. Use Tables 6.8–6.10 to obtain prototype reviews from three external sources. The reviewers should not be members of the team or familiar with the project.

Table 6.6: Evaluation log

Prototype #	Description and purpose	What function is being demonstrated?	What weaknesses can be improved or refined?	What are the next steps?

Table 6.7: Evaluation log

Prototype #	Description and purpose	What function is being demonstrated?	What weaknesses can be improved or refined?	What are the next steps?

Table 6.8: Prototype Review to be filled out by an external reviewer

Based on what the team said, what is the goal of the project?

Based on what the team said, what functions or features does the product need to satisfy?

Describe the strengths and weaknesses of the current prototype

Strengths	Weaknesses

Based on what the team said, how has the prototype been tested against relevant standards or design criteria?

Please list any additional comments or resources the team should investigate.

Table 6.9: Prototype Review to be filled out by an external reviewer
Based on what the team said, what is the goal of the project?
Based on what the team said, what functions or features does the product need to satisfy?

Describe the strengths and weaknesses of the current prototype	
Strengths	Weaknesses

Based on what the team said, how has the prototype been tested against relevant standards or design criteria?
Please list any additional comments or resources the team should investigate.

Table 6.10: Prototype Review to be filled out by an external reviewer
Based on what the team said, what is the goal of the project?
Based on what the team said, what functions or features does the product need to satisfy?
Describe the strengths and weaknesses of the current prototype

Strengths	Weaknesses

Based on what the team said, how has the prototype been tested against relevant standards or design criteria?
Please list any additional comments or resources the team should investigate.

CHAPTER 7

Step 7: Test Solution

Testing is the seventh step of the EDP (Figure 7.1). All prototypes must be tested against established design criteria to identify potential failures or weaknesses of the solution. For instance, a team developing a giraffe feeder may desire that the device last greater than 1 year of use. Upon executing a durability test, the team may identify breakage points of the device that need to be reinforced with stronger material. The team could then revisit Step 6C: Refined Prototyping for advice on refining prototypes.

Testing informs what improvements should be made and is critical for developing a solution that will satisfy both users and clients. Solutions must be tested and found to meet design criteria before they are delivered to the client or potential users.

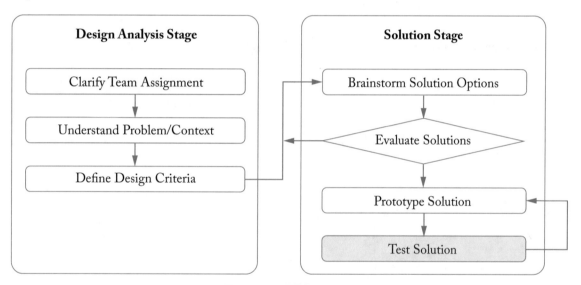

Figure 7.1: EDP: Step 7.

By the end of the section, you should be equipped to:

- Recognize the various test categories and how they can be applied to specific design criteria,

- Identify the method of measurement, procedure, and participants for each test, and

- Evaluate and execute tests to determine whether test results satisfy the design criteria.

Playlist

Watch this video playlist for an overview about testing design solutions. Learn how to conduct appropriate tests for evaluating prototypes against established design criteria.

| Videos | bit.ly/step7-testing |

Video Notes

Define the following key terms.

- Direct test:

- Surrogate test:

- Standard test:

- Constructed test:

What are the four questions that must be answered when setting up tests?

How many tests should a team conduct?

What is the difference between precision and accuracy?

What types of questions can be used to gain user feedback?

Content Summary

The purpose of testing is to evaluate your developed solution against the specified design criteria. There are two main dimensions for categorizing the tests your team conducts.

1. Direct vs. Surrogate: This dimension deals with how measurements are taken. A direct measurement can be taken directly. In other words, a direct measurement is one in which the design criteria itself is measured. However, a surrogate measurement cannot be taken directly; estimation and inference must instead be used. For a surrogate measurement, a substitute measurement of the design criteria is made.

2. Constructed vs. Standard: This dimension deals with how measurements are defined. Standard measurements are preestablished or defined by convention. Constructed measurements, on the other hand, are defined by the team or others (such as the client).

Example tests that span these two dimensions are shown in Figure 7.2.

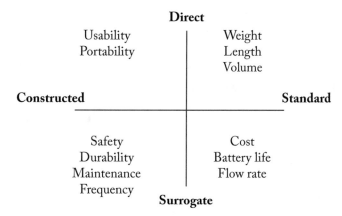

Figure 7.2: Categories of measurement with categorized example tests.

While these test categories differ in type of measurement, they also differ in the number of times the test should be conducted. For instance, a test that is direct may be more precise and require less repetition. Conversely, a surrogate test uses estimation, and thus numerous measurements must be made. A summary of the minimum times a test should be conducted is shown in Figure 7.3.

Figure 7.3: Estimates for number of trials (n) needed.

When creating a testing plan, the following questions should be answered for each test.

- What design criterion is being evaluated?

- How will the measurement be made? For example, what measurement device or method for estimating or calculating will be used?

- How many times will the test be repeated? Justify this number.

- Who will conduct the test and how will the clients or users be involved, if deemed necessary?

Review Questions

1. The main purpose of testing in the EDP is to _____.

    A. Test for failure points

    B. Evaluate your solution against your design criteria

    C. Prove to your client that your solution works

    D. Evaluate your solution against your initial, selected solution idea

2. Which of the following could be a measurement method?

    A. Weighing scale

    B. User-defined scale

    C. Thermometer

    D. All of the above

3. When you are seeking input from potential users, how many users would you ideally speak with?

    A. 1–5

    B. 5–10

    C. 10–20

    D. 20–30

4. Which statement is true about direct/surrogate measurements?

    A. Refers to whether a measurement is direct or indirect

    B. Explains whether the measurement can be made with standard tools

    C. Evaluates the need for estimation or inference for measurements

    D. All of the above

5. Determining how aesthetically pleasing a device is could be an example of _____ testing.

    A. Direct, standard

    B. Surrogate, standard

    C. Surrogate, constructed

    D. Direct, constructed

6. After conducting a test, the results may _____. Select all that apply.

    A. Show that your design successfully meets some design criteria

    B. Show that your design does not meet some design criteria

    C. Be uncertain

    D. Show that the function of the device is consistent with its goal

Exercise #1

Introduction

A design team is tasked with building an improved hay feeder for the giraffes at the local zoo. The team constructs the hay feeder as shown in Figure 7.4. Before deploying the feeder, the team needs to conduct extensive testing. Refer to Step 1: Clarify Team Assignment for the project background and goals.

Task

In Step 3: Define Design Criteria, the team identified food capacity, durability, safety, and aesthetics as key design criteria. Other design criteria include cost and engaging for the giraffes. What is the minimum number of separate tests the team should design?

Develop a testing plan for each criterion in Table 7.1. Include as many details as possible.

Figure 7.4: Hay feeder for giraffes.

Table 7.1: Testing plan for giraffe feeder			
Design Criterion	**How is the measurement made?**	**How many times is the test done? Why?**	**Who is involved in the test?**
Food capacity			
Durability			
Safety			

Exercise #2

Introduction

An engineering team has designed a new dual-purpose backpack. They have described the following tests.

Test 1: Capacity   Target Value: Backpack carries >20 lb

Capacity will be tested by determining how much weight the backpack can support. The backpack will be loaded up with textbooks until it breaks. The weight of the textbooks will be measured with a scale before they are put into the backpack. The test will be completed once.

Test 2: Aesthetically Pleasing   Target Value: >4 on user-defined scale

This criterion will be measured by surveying 20 college students. They will be asked to observe the backpack for 2–3 minutes and then circle one number on Table 7.2.

Table 7.2: UDS for backpack aesthetic				
1	2	3	4	5
Ugly, unattractive	Not pleasing to look at	Neutral, neither ugly or attractive	Pleasing to look at	Very attractive

Test 3: Durability   Target Value: Backpack lasts >5 years

Durability will be tested by estimating how long the backpack will last. A "trial" will be defined as loading a laptop computer and three textbooks into the backpack, zipping and unzipping three compartments, walking with the backpack for 5 minutes, and then dropping it on the ground. The trial procedure will be conducted 300 times. A team member will conduct this test.

Task

What are some strengths in each of the three tests? Use direct quotes when applicable.

How can these tests be improved? Identify any errors or gaps. Use direct quotes when applicable.

Applying This Step to Your Project

Task: Recall your Design Criteria

Begin by copying your design criteria from Step 3: Define Design Criteria onto Table 7.3 and Table 7.4. Review all of the established objectives and constraints, as well as target values. Discuss how each of these design criteria were defined.

Task: Decide on a Testing Strategy for Each Design Criterion

For each design criterion, consider how one or more tests could evaluate whether that design criterion is met. In other words, generate ideas for tests that could verify your stated design criteria. Verification can be accomplished through tests or calculations; however, testing is preferred.

For criteria that involve physical properties, you may need to conduct tests that probe these properties. For example, capacity is a physical property of a solution that can be tested by measuring the volume or weight. Design criteria that involve feelings or opinions may require a collection of responses from people. Design criteria that are for abstract characteristics may need comments from experts or indirect verification. For example, safety can be verified by soliciting the opinions of experts.

Task: Carefully Specify Each Test

For each developed test, you should answer these four questions on Table 7.3 and Table 7.4.

1. What design criterion is being measured? What is the target value?

2. How will the measurement be made? For example, what measurement device or method for estimating or calculating will be used?

3. How many times will the test be repeated? Justify this number.

4. Who will conduct the test and how will the clients or users be involved, if deemed necessary?

In the creation of your testing plan, be specific and numerical with each verification method. For example, durability can be tested as "the device will survive a drop test from 5 feet for 50 times and still sustain the critical functions of X, Y, and Z." The test description would also need to include specifics on the type of surface onto which the device is dropped, what range of orientations that it is dropped, how the functions of X, Y, and Z are measured, who will do the tests, etc.

Task: Conduct Tests for Design Criteria

Use your testing plan to conduct detailed tests for each of the design criteria. Verify whether the test meets the target value or specify where design changes may need to be made. You can record the test data and conclusions on Table 7.5 and Table 7.6. You may need to consult Step 6C: Refined Prototyping for refining the prototype if it does not satisfy all design criteria during testing.

Table 7.3: Blank testing plan

Design Criterion and Target Value	Circle the Categories of the Test	How will the measurement be made?	How many times will the test be repeated?	Who will conduct the test? Will client/users be involved?
	Direct Surrogate Standard Constructed			
	Direct Surrogate Standard Constructed			
	Direct Surrogate Standard Constructed			

Table 7.4: Blank testing plan

Design Criterion and Target Value	Circle the Categories of the Test	How will the measurement be made?	How many times will the test be repeated?	Who will conduct the test? Will client/ users be involved?
	Direct Surrogate Standard Constructed			
	Direct Surrogate Standard Constructed			
	Direct Surrogate Standard Constructed			

Table 7.5: Testing log

Date	Design Criterion and Target Value	Test Results (Write in data or provide source link.)	Did the test pass or fail?

Table 7.6: Testing log

Date	Design Criterion and Target Value	Test Results (Write in data or provide source link.)	Did the test pass or fail?

CHAPTER 8

Step 8: Finalize a Solution

Engineers develop solutions to complex problems through the iterative process of research, design, and testing. After analyzing a problem, an individual or team can develop a conceptual solution into a concrete product. This design process is not linear and many steps may need to be repeated or revisited. Nonetheless, each design project should proceed through all steps of the design process outlined in Figure 8.1.

The ultimate goal of the design process is to create a product or program that solves the problem at hand. Sometimes the process may be stopped before a solution is finalized. In this case, the project can be continued by another engineer or used for educational purposes. Even if a "final" product is not developed, the EDP is a robust framework that allows for a problem to be critically analyzed. It allows both teams and individuals to create and develop potential solutions.

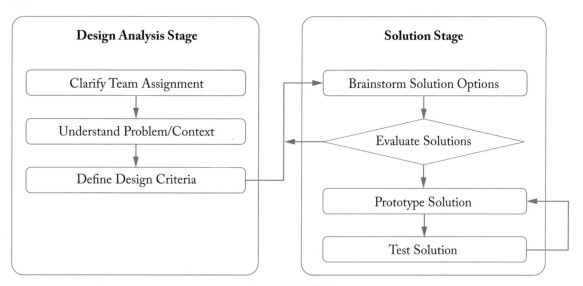

Figure 8.1: EDP.

By the end of this workbook, you should be equipped to:

- Define a design problem that needs to be solved,

- Conduct research to better understand the scope of a problem,

- Create quantitative design objectives and constraints,

- Brainstorm solutions for multiple design block,

- Evaluate solution ideas against desired criteria,

- Develop prototypes to demonstrate design functions, and

- Test prototypes to determine areas of improvement.

Content Summary

Each project should conclude once the initial goals have been reached. While some projects end when time runs out or resources are no longer available, other projects end because the problem has been solved and the product is ready to be delivered to a client. To determine whether a project has reached a stopping point, the following questions can be answered in Figure 8.2.

- Does the existing prototype or solution satisfy all the design criteria?

- Is there a client for this project?

- Do you have the time, materials, and resources to continue the project?

- Can the project be completed by someone else?

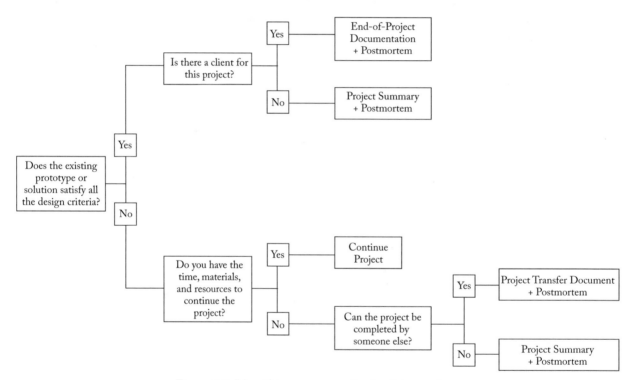

Figure 8.2: Identifying next steps for finalizing project.

For instance, a project that has not satisfied all the design criteria and no longer has time, materials, or resources may be transferred to a new engineering team. The new team could then finish the project through additional prototyping and testing against design criteria.

Any finished project requires sufficient documentation. The type of documentation depends on the final destination of the project and the user. For example, if you are designing a new eating device for children with musculoskeletal disorders, then you may need to write an instruction manual for the children and parents. If this device will be used in a hospital, you may need additional text on how to repair the device, how it is assembled, and how much each component part costs. Projects that are sponsored may have additional documentation requested by the client. Some common forms of project documentation are explained in Table 8.1. Engineers should keep copies of all documantation for their personal portfolios.

Table 8.1: Types of end-of-project documentation		
Name	**Purpose**	**Who is the audience for this document?**
Construction details	Summarizes all technical specifications for the project. Can include building instructions, bill of materials, design drawings, CAD, electronic schematics, plan for manufacturing, etc.	Client, professional manufacturer, engineering team
Instruction manual	Provides information on how to assemble or disassemble the product. Explains how to safely use the product.	Product users
Provisional patent draft	Identifies similar patents and novel characteristics of your product. Begins the process of obtaining a patent for a unique invention.	Client, engineering team, or designer
Standards and regulations	Identifies necessary laws that the device must satisfy, often in terms of safety and testing. This document is particularly important for medical devices or solutions with user interaction.	Client, engineering team or designer, government entity
Project transfer document	Provides sufficient technical information on a project that another team or individual could continue the design. Often includes similar information as construction details. Identifies strengths, limitations, and future work for the project.	Future engineering team or designer

Regardless of what type of final documentation is needed for a project, all projects should conclude with a postmortem. The postmortem is a finishing activity that reviews the process, decisions, successes, and limitations of the teaming throughout the life of the project. It is critically important for designers, even those not working in teams, to hold a postmortem for reflection and growth.

Exercise #1

Task

Read the following project status descriptions. Determine potential next steps for the project. Be sure to identify at least one type of final documentation that is needed.

1. A team has developed a product that satisfies four of the five design criteria. The client has not granted an extension for the project.

2. A designer has completed a novel personal project and hopes to sell the invention.

3. An engineering team has finished designing a device that satisfies all design criteria and is being handed off to a manufacturing team.

4. A design team has completed a project for a client and handed off the device. The client is a physical therapist and plans to use the device in the clinic.

5. One division within an engineering company designs a product to the limit of their expertise. They are handing the design off to another team with greater expertise.

Applying This Step to Your Project

After completing the decision diagram in Figure 8.2, complete the tasks that correspond to your project's status.

Task: Continue Project to Completion

Identify concrete goals for continuing the project. Does the prototype need to undergo further testing against the design criteria? Does a component need to be repaired or redesigned? Does the client have additional wishes that need to be considered? Do successive prototypes need to be completed?

After establishing goals and considering what resources are available, develop a new project timeline. You should identify immediate next steps as well as detailed future plans. More information on project planning is in Appendix D: Project Planning. Additional resources for continuing a project include the following.

- *Product Design and Development,* by Karl Ulrich and Steven D. Eppinger, and

- *Making It: Manufacturing Techniques for Product Design*, by Chris Lefteri.

Task: Project Video

Reflect on the most recently completed prototype. Record a video to describe and demonstrate the final design solution. The video could include the following.

- Brief statement of problem and its importance.

- Full explanation and detailed description of the final solution. Discuss important design blocks and components, as well as key functions.

- Demonstration of the solution "in action". While in operation, narrate and explain how it is operated.

- Short discussion of whether and how the prototype meets the set design criteria.

- Short list of tasks remaining for completion of the project, if necessary.

Remember to discuss important technical details, such as size, material, specially selected components, etc. You should discuss the independent design blocks and how parts are integrated. For projects involving electronics, you should explain the software, hardware, and program functions.

Task: Project Summary

Complete the Project Summary page to document key information. Attach pictures of the prototype (including close-up images of components) to the summary page, if you do not make a video.

Task: Create End-of-Project Documentation

Review Table 8.1 to determine which final documents are relevant to your project. Compile all drawings, schematics, flow charts, and electronic files (e.g., CAD) for the prototypes if you are making a construction manual. Make a list of parts, both bought off-the-shelf and constructed from scratch, to help with the summary. You may need a detailed description of how to assemble the device (for provisional patent) or a simplified version with many images (for instruction manual). Create these documents in an electronic format.

Task: Create Project Transfer Document

A document summarizing the project is helpful for any team or individual that may be continuing work on the problem at a later time. While the project video and end documentation provide technical details, the Project Transfer Document provides information on critical design decisions. It can inform future teams on why a particular solution was selected, why certain criteria were prioritized, or even what design limitations exist. Be sure to include a link to all source files or attach the following documents.

- Client interview notes

- Background research

- Rationale for design criteria

- All decision matrices (pairwise comparison chart, morph chart, screening and scoring Pugh matrices, work breakdown structure)

- Images of successive prototypes

Task: Conduct Postmortem

If you are not continuing with the project, conduct a formal postmortem. The postmortem is a reflection on the skills you learned, as well as strengths and weaknesses of the team and prototypes, and the design process. While other types of end documentation focus on the quality and technical specifications of the product itself, the postmortem focuses on organization, communication, decision making, and adaptability during the design process. See Appendix B, Section B3: Team Postmortem.

Project Summary

Client Name	Email	Phone Number

Team Member	Email	Phone Number

Problem statement:

Current state of the prototype:

Successes	Limitations

Three areas of improvement or critical functions that should be evaluated:

1.

2.

3.

Link to project summary video or folder with images:

Link to online project folder with all source files:

Project Transfer Document

Client Name	Email	Phone Number

Team Member	Email	Phone Number

Problem statement:

Ranked Design Criteria and Target Value	Pass or Fail?	Comments

Comments on testing plan (e.g., number of samples, important procedures):

Current state of the prototype:

Successes	Limitations

Suggested future work, including needed prototyping and testing:

1.

2.

3.

4.

5.

Link to project summary video or folder with images:

Link to online project folder with all source files:

Project Terminology	
Term	**Definition**

APPENDIX A

Using this Workbook

This workbook and related materials are designed for learners who are in high school and undergraduate programs. Sections of the materials can be used for students as young as those in middle school and as old as those in graduate school. For individuals outside of formal education who have project ideas, like hobbyists, designers, and makers of any age, this workbook's processes can be beneficial. To date, the work has been piloted, tested, and refined extensively for use in first-year engineering courses and upper-level high school engineering courses. It has been deployed in both high- and low-income settings in the U.S. and abroad.

The materials and workbook can be used inside or outside of a formal educational structure, although some formal structure is suggested. The materials and workbook were developed for teams of individuals working together on a shared problem, but the work can also can be completed by an individual.

Appendix A gives additional instructions and context about how to approach learning from and using these materials.

Flipped Model

This workbook relies on a flipped classroom model, which was informed by Bloom's taxonomy and best practices in engineering education. At a high level, learners grasp the EDP and complete short exercises to test their knowledge. Then, individuals apply their knowledge to their own unique problem. It is suggested to use the materials in the order shown in Figure A.1.

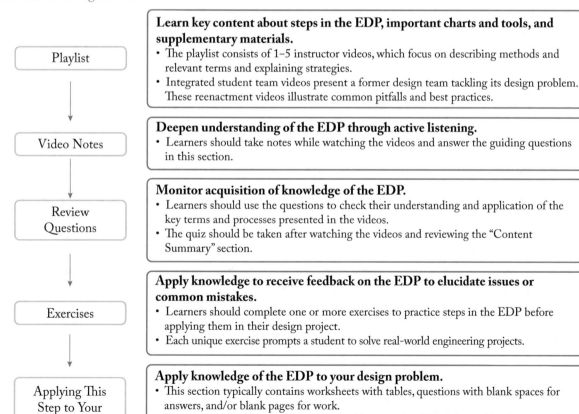

Figure A.1: Order of learning and applying EDP materials.

When using these materials in a *formal* educational setting, the first three sections should be completed <u>prior</u> to coming to class. <u>During</u> class, the students should complete "Exercises" and "Applying This Step to your Project." In a class, both instructor and peer support are available, and students can complete the cognitively challenging tasks of the EDP.

Timeline Suggestions

The use of the workbook and completion of a design project can occur over time frames that range from several months to several years. The overarching structure of the workbook is the application of the EDP to solve a problem with these materials in order (Figure A.2), given the caveat that repeating a step may be necessary in order to solve a problem, as iteration is a key component of the EDP.

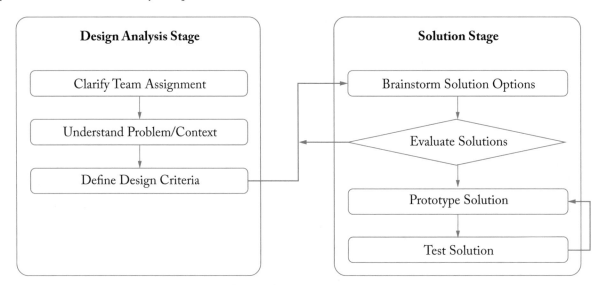

Figure A.2: Order of the engineering design process.

The appendices of this workbook contain additional video modules and exercises. They supplement the design process and provide content for professional development.

Appendix B: Teaming

- The team orientation (Section B1: Team Orientation) helps teams understand their peers.

- A team pit stop (Section B2: Team Pit Stop) monitors the team's dynamic to mitigate conflicts and inefficiencies.

- The project postmortem (Section B3: Project Postmortem) provides a formal questionnaire for learners to reflect on the strengths and weaknesses of their team and help prepare learners for future collaborative projects.

Appendix C: Communication

- An oral presentation (Section C1: Oral Presentation) provides an opportunity for learners to summarize their progress and articulate their design decisions.

- A technical memo (Section C2: Technical Memos) helps students articulate their work and document the specifications or details of the solution process.

Appendix D: Project Planning

- Tools used in project planning (Section D1: Work Breakdown Structure and Section D2: Planning Tools) help teams stay on track to meet set deadlines. This can be used at the outset of a project, but certainly by the prototyping period.

Below are some examples of prior uses and can serve as a guide.

One-semester course at college or university (approximately 15 weeks)

- The first half of the semester is devoted to clarifying the design problem, developing the design context review, establishing design criteria, brainstorming solutions, using a Pugh matrix to evaluate and select a solution, and then defining the selected solution in detail. Each listed step lasts 1–2 weeks.

- The second half of the semester is devoted to student teams iteratively building and testing a series of prototypes.

- Supplementary materials, including project planning, communication, team building, etc. can be interspersed when appropriate.

Two-semester course at college or university (approximately 30 weeks)

- The first ~10 weeks is devoted to clarifying the design problem, developing the design context review, establishing design criteria, brainstorming solutions, using a Pugh matrix to evaluate and select a solution, and then defining the selected solution in detail. Each listed step lasts ~2 weeks.

- The remaining ~20 weeks is devoted to student teams iteratively building and testing a series of prototypes.

- Supplementary materials, including project planning, communication, team building, etc. can be interspersed when appropriate.

Supporting Roles in a Formal Structure

Throughout the use of this workbook in a formal course, teams benefit from regular contact with an instructional support team, which includes course faculty, technical mentors (those with relevant technical expertise), teaching assistants, writing consultants, and shop technicians. These individuals coach the design teams by sharing their expertise, modeling more rigorous, systematic thinking, and providing formative feedback. This type of guided discovery or cooperative learning allows instruction and feedback to be tailored to the needs of individual students and teams. It also mitigates the cognitive overload many students experience when they simultaneously attempt to acquire new knowledge and engage in complex problem solving. Depending on the structure of the course and its budget, some or all of the following can be engaged.

Professional Support

- *Technical mentors* are selected based on the needs of the projects and technical gaps in students' knowledge. For example, a team designing a flow controller for a water purification system may be assigned to a faculty member of the mechanical engineering department who can provide technical insight for fluid dynamics and controller design. Students are expected to meet weekly with their mentor to receive the most support.

- *Shop technicians or engineers* are employed by the design space itself and assist students when using complex machinery and processes. Student lab technicians are also employed to aid peers when the formal technicians are not available, such as outside regular business hours. Having people skilled with the machines and prototyping tools at the design facility ensures students are learning proper technique and avoiding potential injuries.

Peer Support

- *Teaching assistants* are students who have previously taken the course, so they are well versed in the EDP and expectations of the class. They are typically assigned to several teams and can stay abreast with the teams' progress to provide immediate, relevant feedback.

- *Writing consultants* are also students who have previously taken the course and demonstrated high proficiency with technical documentation. In a university setting, students may be given writing assignments aligned to each step of the process. The goal of these assignments is to learn technical writing skills and how to justify design decisions. Writing consultants work individually with teams to provide feedback and model technical writing best practices.

Supporting Roles in an Informal Structure

For those who are using this workbook piecemeal or as a guide for their own project, regular contact with experts or others can assist in a number of ways to improve the ultimate project. In these situations, the traditional roles of instructional support and subject matter experts shift to more informal roles found from close contacts or community members in person or online. Consider finding a mentor who you can use to check in about the project, ask questions, or keep to a schedule with. When you are stuck, ask this mentor or look for someone who might know the answer to ask questions. It is never a bad practice to learn how to reach out to experts and ask them for counsel. While this workbook has been most effective in a structured capacity, it can also be effective when used informally. Some of the roles that exist in the structured format shift to processes or communities of others online to support and contribute feedback to the project.

Professional Support

- *Project mentors or subject matter experts* function to help the team/individual succeed based on the needs of the projects and technical gaps in the knowledge of those working on the project. For example, a team/individual designing a solar powered go-kart would look for an expert in their surrounding community or online who is in the automotive industry or an electrical engineer who can provide technical insight for the power systems and electrical systems. Mentors are best utilized when they are regularly kept up to date on the project.

- *Prototyping experts* are individuals who work at a makerspace or have prototyping facilities that they have access to. These individuals may be able to provide guidance or access to these tools to help the project prototype develop. Working alongside people skilled with the machines and prototyping tools ensures that proper techniques are learned, which can help avoid potential injuries.

Peer Support

- *Online communities* can be extremely useful for getting questions answered or discovering new ways to move forward in your projects. These online communities are usually populated by other passionate individuals who can help you through the struggle of working through a difficult project. Engineering. com and Edisonnation.com are two examples of forum-based sites where you can receive assistance on your project from like-minded individuals.

- *Project documentation* sites are online locations where individuals/teams can document the work that they are doing to receive feedback and complete the best practice of documenting their project regularly.

There is some overlap with documentation sites and online communities, however these documentation sites are excellent locations to provide current and final documentation for others to learn what you are doing and replicate your project. Examples of these locations are instructables.com, hackster.io, and makeprojects.com.

APPENDIX B

Teaming

Oftentimes in the field of engineering, projects are too large to be carried out by a single individual. Rather, long-term projects benefit from a diverse team of people, with various strengths and weaknesses, working to find a solution. This appendix serves as a guide for defining effective teamwork and improving overall team performance. The three substeps contained here should be used at various points throughout the design process.

Although teams may differ in their values and work ethic, high-performing teams display common characteristics, as described in Section B1: Team Orientation. Learning about high-performing teams is typically done at the onset of a project. Section B2: Team Pit Stop outlines how to conduct a pit stop, evaluate team effectiveness, and remedy any conflicts. A pit stop can be done any time during a project and may be repeated. Section B3: Team Postmortem provides information for conducting a postmortem, which allows individuals and teams to reflect on the design process. Postmortems are typically done when a project is complete.

SECTION B1: TEAM ORIENTATION

Playlist

Watch this video playlist for an overview about teamwork. Learn best practices that high-performing teams use and some characteristics of low-performing teams.

Videos	bit.ly/additionalstepsb1-teamorientation

Content Summary

There are typically several roles on a team. The *leader* creates an agenda for each meeting and keeps the team on task. The *facilitator* ensures that every team member participates and contributes evenly. The *organizer* keeps track of materials and ensures that due dates are met in a timely manner. The *scribe* records and distributes information to the team.

In addition to assigning roles, it is important to recognize the difference between team meetings and work sessions. Work sessions do not require the entire team and are usually longer than an hour. A meeting, however, typically requires the entire team and ideally lasts less than 1 hour. An example agenda for a team meeting is shown below in Figure B.1.

Meeting Information			
Date:	Tues, 04/14/2015	Location:	OEDK Conference Room
Start Time:	7:00pm	End Time:	8:00pm
Leader:	Marie	Facilitator:	Josiah
Organizer:	Sarah	Scribe:	Kayla
Action Items from Previous Meeting			
Report on visit and discussion with client			Josiah
Demonstrate new soap grater			Kayla
Agenda Items		**Presenter**	**Time Alloted**
Discuss soap grater options from various kitchen retailers		Kayla	10 minutes
Plan and delegate roles for poster presentation		Marie	15 minutes
Present issues with matching process		Josiah	15 minutes
New Action Items		**Responsible**	**Due Date**
Test grater efficiencies		Sarah and Kayla	04/16/2015
Draft Poster		Marie and Josiah	04/17/2015
Other Notes or Information			
Next team meeting—Monday, April 20, 2015, 5:30 pm, Coffeehouse in the RMC			
Poster competition—Friday, April 24, 2015, 1:00 pm, Tudor Fieldhouse			

Figure B.1: Example agenda for a team meeting.

High-performing teams exhibit the characteristics shown in Figure B.2. In contrast, low-performing teams may lack these characteristics, which can cause internal conflicts and slow project progress. For instance, a team that does not have good communication may prototype independently without relaying their progress to other team members.

Figure B.2: Characteristics of high-performing teams.

Review Questions

1. Who makes sure that everyone is contributing to the team?

    A. Leader

    B. Organizer

    C. Facilitator

    D. Scribe

2. Important activities during team meetings include which of the following? Select all that apply.

    A. Generate ideas

    B. Report on progress

    C. Prototype

    D. Plan

3. Agendas should not identify which of the following?

    A. Start time and end time

    B. Location

    C. Member's tasks

    D. Anticipated action items

4. Which of the following help make team meetings the best they can be? Select all that apply.

    A. Having an agenda

    B. Having flexible end times

    C. Assigning roles

    D. Planning successive meetings

5. Self-assessment is important for teams because _____.

    A. It helps them critique personal performance

    B. It helps them determine whether they are meeting team goals

    C. It helps them identify areas that need improvement

    D. All of the above

6. Your team has a member who is decidedly less dedicated and doing less work than other members. What is a good option for your team to do?

    A. Delegate the student's tasks to other members so that you don't fall behind.

    B. Ask the student to give the student's spot on the team to another university student who is more willing to contribute.

    C. Hold a meeting with that member to figure out how the student can more strongly contribute to the team.

    D. None of the above.

7. When you are committed to a team, you should be _____.

    A. Dedicated

    B. Willing to change

    C. Empathetic

    D. All of the above

Exercise #1

Introduction

Below are descriptions of some basic values that human beings hold, presented on a continuum, adapted from work done by Hofstede. When we assume our colleagues share our values and perspectives, misunderstandings often can occur. Teams will work together more effectively if team members can anticipate ways their differences might derail their communication and interactions. This exercise can be helpful during team formation.

Task: Individual Response

Quickly read the descriptions and check the box that best reflects your position on the continuum for each value.

Individualism (importance of self)	Collectivism (importance of group)
While you may seek input from others, you are ultimately responsible for your own decisions regarding where you live, what your major is, or your career choice. You have a sense of pride in being responsible for yourself and know that others expect you to be independent.	You make important life decisions based on the needs of the group and put the well-being of the group ahead of your own. You make major life decisions in consultation with your family, friends, and co-workers. Identity is a function of one's membership or role in a primary group.

High Individualism ❑   ❑   ❑   ❑   ❑   ❑   ❑   High Collectivism

Polychronic Time (unlimited)	Monochronic Time (commodity)
You feel that time is an unlimited good and available as needed. People should take the amount of time necessary to do what they need to do. Life does not follow a clock; things happen when they are supposed to happen. Deadlines can be changed and plans are fluid.	You feel that time is a precious good and should not be wasted. Human activities, as a result, must be carefully organized. You take great care to plan your day and to arrive on time to class and meetings with friends and family, whose precious time you feel you must not waste by being late.

High Polychronism ❑   ❑   ❑   ❑   ❑   ❑   ❑   High Monochronism

Change/Progress/Risk-Taking	Stability/Tradition/Risk-Aversion
You believe that almost everything around you will change—even the friends you have throughout your lifetime. You believe that change bring many positives to your life and means progress. As a result, you are very willing to try to new things and ideas.	You recognize it's important to keep traditions because they bring an expected rhythm to life. Stability gives meaning to life, which change disrupts. You prefer to do things as you or others have done them in the past.

High Change ❑   ❑   ❑   ❑   ❑   ❑   ❑   High Stability

Indirectness	Directness
You believe indirect communication is the best way to respect others' feelings and integrity. If you have a problem with another person, you might leave subtle clues that there is a problem. A face-to-face confrontation would be seen as rude and offensive. This approach maintains the harmony of the community.	You like it when people say what they mean and mean what they say, and there is no need to read between the lines. If there are problems, you like to have face-to-face conversations to resolve them. Giving information efficiently is more important than saving someone's feelings.

High Indirectness   ❑   ❑   ❑   ❑   ❑   ❑   ❑   High Directness

Equality	Hierarchy
You believe that people should interact with each other on a level playing field. While differences such as age and education obviously exist, you don't feel these should be the sole basis for interacting with others. People should be judged on merit, and everyone has a right to contribute.	You believe strongly that people should be treated according to their status. The views of more experienced or older individuals should be respected and given priority. Hierarchy is a fact of life and gives everyone a sense of their place in the world.

High Equality   ❑   ❑   ❑   ❑   ❑   ❑   ❑   High Hierarchy

Adapted by the Rice Center for Written, Oral, & Visual Communication (CWOVC) from the "Core Cultural Values and Culture Mapping" activity developed at the University of Minnesota and the "Cultural Dimensions and Communication" worksheet developed by Beth O'Sullivan of the Rice Jones School and RCEL (based on the work of Craig Sorti).

Task: Team Response

Now, using the individual value worksheets each team member completed, use this sheet to aggregate your responses. You should use only ONE COPY PER TEAM. Make one hash mark per person on each line below, representing the location of each team member on that continuum. Place the team member's initials below that hash mark.

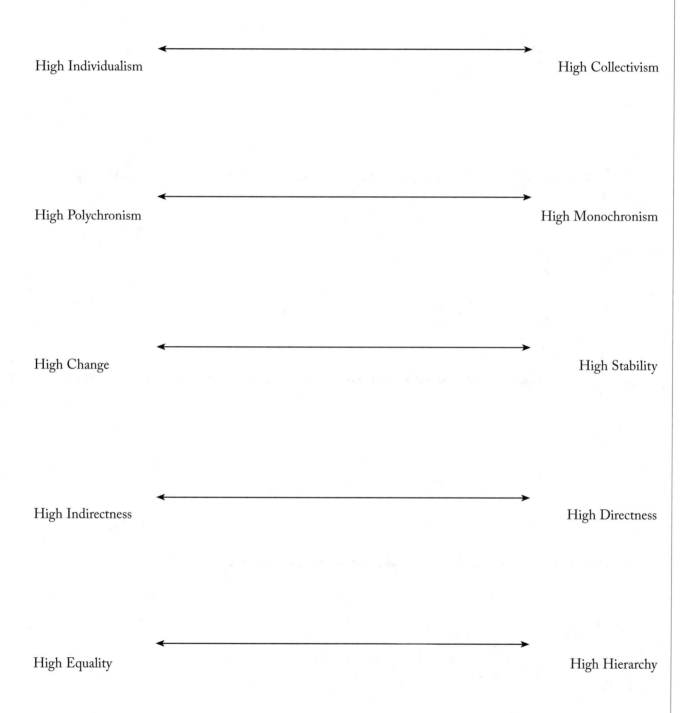

High Individualism   High Collectivism

High Polychronism   High Monochronism

High Change   High Stability

High Indirectness   High Directness

High Equality   High Hierarchy

Task: Reflection

Consider your team's diversity and the varying approaches of the members of your team.

1. What are the benefits or advantages of working in a team that includes members with different backgrounds, values, and beliefs?

2. What differences do you see among yourselves? What issues could potentially arise from these differences?

3. How can you start addressing these differences now so that the benefits and advantages can be realized?

4. If you are alike in several dimensions, how could this be beneficial or detrimental?

SECTION B2: TEAM PIT STOP

Content Summary

A pit stop is an opportunity for a team to reflect on their performance, identify any issues that are hindering progress, and create a plan to address the issues moving forward. It is important for each member to reflect independently before reviewing responses as a team. Assigning roles for the pit stop activity, such as scribe and facilitator, can also be helpful.

The pit stop is a time for teams to be honest about their shortcomings in a respectful and productive way. Several best practices for constructive communication include the following.

- <u>Problem-oriented language:</u> Critiques should focus on problems, issues, or behaviors that can be changed (e.g., frequency of communication, time management). When possible, the critiques should not target specific team members.

- <u>Descriptive language:</u> The team should describe what happened, explain the reaction or consequences, and suggest a solution. Specific events or behaviors should be addressed, not general problems. For example, one specific problem would be that a team missed a deadline for an assignment. A general problem would be saying that a team is always late.

When brainstorming solutions to any team problems, there should be a concrete plan. It is vital to not only propose a solution, but also to define what steps need to be taken and how success will be measured in the future. An example of this thought process is shown in Table B.1. Examples, exercises, and best practices were developed by David Nino (MIT).

Table B.1: Example implementation plan for team problem	
Problem	Meetings often start 15-20 minutes late because two or more team members are late.
Proposed Solution	Plan meetings in advance and set reminders.
Steps to Take (Execution)	1. Meetings will be scheduled during class so all team members can agree on the time. 2. All team members will set reminder alarms 3 hours and 15 minutes prior to a meeting.
Measure of Success	All future meetings start within 5 minutes of scheduled time. The team will reevaluate this implementation plan after the next five team meetings.

Exercise #1

Task: Individual Response

Answer the questions listed below.

- What are your personal hopes and aspirations for the group? List your aspirations.

- What group factors are contributing positively to the achievement of these hopes and aspirations? List three team strengths.

- What group factors are hindering achievement of these hopes and aspirations? List three team weaknesses.

Task: Team Response

Review the individual responses in the team and identify common themes.

- Discuss above notes with your design team. Try to achieve some consensus on the team's aspirations, strengths, and weaknesses. Write down key points.

Task: Develop Implementation Plan

Answer the following questions in the team. Create a plan for implementing changes to address important issues.

- What are the two or three most important issues that we need to address?

- What actions will make a real difference in our performance?

- How will our team implement these changes now?

- How will the team know that the changes are making a difference?

Task: Develop Implementation Plan

SECTION B3: TEAM POSTMORTEM

Content Summary

A postmortem is an opportunity for teams to learn from their experiences and carry insights into future teamwork. Lessons learned can be positive, reinforcing behaviors for effective teaming, or negative, surfacing issues and behaviors that should be avoided. Another purpose of a postmortem is to give closure to a project. Closure is important for team members who are breaking away and moving to different projects, or to wrap up a particularly long or tough project.

Conducting this type of reflection requires teams to be honest and constructive. Teams should start from the beginning and walk through the overall timeline of the project. It is important to discuss one topic at a time, provide specific examples, and ask many questions. The following questions should be addressed for each topic.

1. What were our original aspirations and goals?

2. What did not go so well? Why?

3. What went well? Why?

4. What principles should we recommend for future teamwork?

Examples, exercises, and best practices were developed by David Nino (MIT).

Exercise #1

Task

Conduct a postmortem and record the findings. Here are potential topics to discuss.

Planning

- Were the group goals clear?

- How well did the team plan throughout the semester?

- How could planning be improved?

Scheduling

- Was the schedule realistic? Was the schedule detailed enough?

- Looking over the schedule, which tasks could have been better estimated?

- What were the biggest obstacles to achieving the scheduled dates?

- How was project progress measured? Was this method sufficient?

- Was the project schedule maintained throughout the project? How were changes managed?

Communication

- Were there any issues with communication inside your group?

- Were there any issues with communication between your group and other groups or individuals?

- Was the team lead effective in communicating information?

- How were conflicts or disagreements handled? Were they managed well? Poorly?

- Were there any issues that were left unaddressed?

Team/Organization

- Was the role of each member clear? Did the roles evolve over time?

- Were there any issues with teamwork or morale?

- Was workload shared appropriately?

- How did you maintain accountability of work and deliverables?

- Did the team work together well? Why or why not?

Management

- How were decisions made within the team?

- Did you establish a decision-making process or set of rules?

- Were decisions communicated effectively to the team?

- Did you have any changes of scope? Were scope changes effectively communicated?

Quality

- Are you happy with the quality of the final product?

- Could the final product be done better? How?

Summary

- Overall, list items that went well.

- Overall, list items that need to be improved. Suggest how to improve.

- List any additional issues that should be discussed.

APPENDIX   C

Communication

In a world of advancing technology and global cooperation, communication is vital. Successful engineers can effectively communicate their ideas to diverse audiences. Through intentional communication, engineers can demonstrate their mastery of the content and increase credibility in their designs. This appendix serves as a guide for communicating design ideas in both oral and written formats.

Section C1: Oral Presentation provides guidelines for oral presentations. This section examines characteristics of effective technical presentations and includes advice for oral delivery. Often, a design solution should be described with more detail than is feasible during a short, oral presentation. Technical memos satisfy this need to convey key information concisely in an executive summary format. The following Section C2: Technical Memos outlines how to write a technical memo capturing decisions made during the design process.

SECTION C1: ORAL PRESENTATION

Playlist

Watch this video playlist for an overview about presenting a design proposal. Learn how to deliver a technical presentation effectively. Images, examples, exercises, and best practices were developed by Dr. Tracy Volz (Rice University).

Videos	bit.ly/additionalstepsc1-oralpresentation

Video Notes

What five behaviors do high impact presenters exhibit?

What is the difference between inductive and deductive presentations?

List three methods for establishing credibility in a presentation.

What is the best font type for PowerPoint presentations?

Content Summary

Effective communication is vital for success as an engineer. When giving an oral technical presentation, it is beneficial to use a deductive structure. In a deductive presentation, the thesis or recommendation is given at the beginning. The presentation then expounds on the background of the problem, relevant data, and other key details. An example for a final design presentation with a deductive structure is shown in Figure C.1.

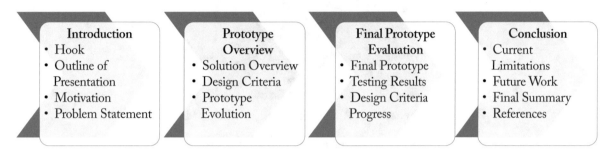

Figure C.1: General outline for technical presentations.

Technical presentations also benefit from clear visual aids. Slide text should be succinct and consistent, allowing the reader to grasp the main ideas without reading through blocks of text. Figure C.2 shows a poorly made slide on the left and its improved counterpart on the right.

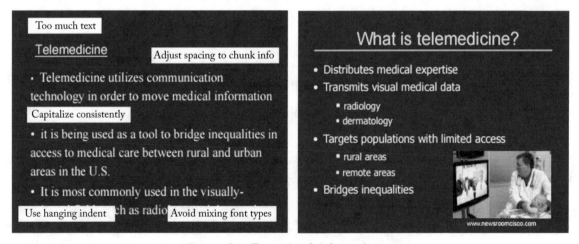

Figure C.2: Example of slides with text.

Before incorporating visuals, ensure that they are relevant to the thesis and high quality. Visuals should also be scaled, captioned, and labeled. This is especially important for complex or intricate prototypes. All prototypes should be photographed with a clear background to minimize clutter and distraction, whenever possible. An example of effective labeling of a prototype is shown in Figure C.3.

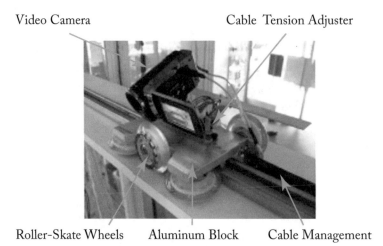

Figure C.3: Example of a prototype with labels.

When presenting data in a graphical format, remove unnecessary information, choose large and readable fonts, add labels, and use color carefully. Common pitfalls when presenting data include:

- Having nondescriptive or nonexistent titles,

- Having small, unreadable font,

- Using colors with low contrast,

- Showing large amounts of data with no visual cue,

- Using highly technical language with no explanation, and

- Repeating information in several areas of the graphic.

Figures C.4–C.6 help elucidate these common mistakes when presenting data. In each figure, the poorly designed slide is displayed on the left and the improved slide is on the right.

Telomerase Activity Assay Ranks #1							
Evaluation Criteria	Weight (%)	Telomerase Activity Assay of Sputum		Spiral CT		PET	
		Rating	Weighted Score	Rating	Weighted Score	Rating	Weighted Score
Early Detection	18	4	0.72	3	0.54	3	0.54
Cost	16.5	4	0.66	3	0.495	1	0.165
Sensitivity	15	3	0.45	4	0.6	2	0.3
Availability	12.5	4	0.5	3	0.375	2	0.25
Safety	12	4	0.48	3	0.36	2	0.24
Specificity	10	4	0.4	3	0.3	4	0.4
Non-Invasiveness	8	4	0.32	5	0.4	3	0.24
Response Time	8	3	0.24	3	0.24	3	0.24
Total Weighted Score			3.77		3.31		2.375

Figure C.4: Example of slides with matrices.

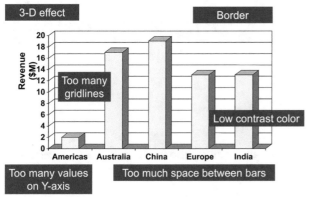

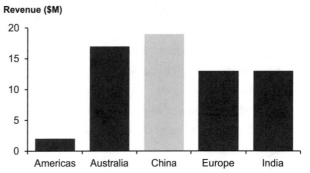

Figure C.5: Example of slides with bar graphs.

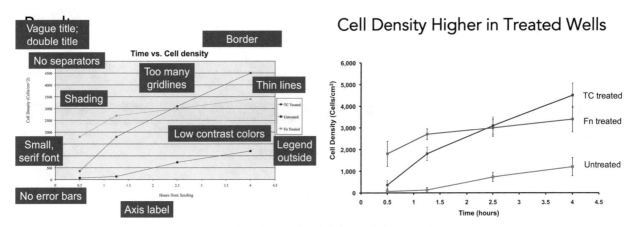

Figure C.6: Example of slides with line graphs.

Table C.1 presents a checklist that can be used for delivery, visuals, and questions and answers.

Table C.1: Oral presentation checklist*
Delivery
• Convey enthusiasm
• Avoid fillers ("you know" and "ums")
• Speak loudly enough to be heard
• Pronounce words clearly
• Speak without noticeable pauses mid-sentence
• Use falling intonation at the end of sentences (avoid upward "questioning" intonation that seems uncertain)
• Vary pace
• Use inflection for emphasis
• Maintain eye contact with audience, especially during introduction and conclusion
• Look at all portions of audience
• Use concept-related gestures
• Avoid nervous hand movements
• Stand straight
• Plant feet or move with purpose
Visuals
• Informative slide titles
• Appropriate number of words and items on slides (not overloaded)
• Font sizes can be read easily
• Diagrams focus on critical elements, not excessive in detail
• "Detail" or "excerpt" diagrams identify key components or decisions
• Comparisons that make analysis or design choices clear and easy to evaluate
• Background that does not interfere with diagrams or text
• Text and diagrams are sufficiently far from edges of slide to make organization of information clear through "white" or empty space
• Images appropriate for audience
• Color choices are easy to view
Questions and Answers
• Listen to question without interrupting
• Repeat or paraphrase question
• Address response to everyone not just person who asks question
• Appear open and confident
• Step toward questioner or hold position
• Begin answers with general statement of overall answer to question, and then add details or explanation
• Do not evaluate questions ("That's a good question.")
• Challenge definitions or criteria presented in questions that are not relevant
• Wrap up well

*Prepared by Tracy Volz, Ph.D.   (Rice University)

Review Questions

1. Which of the following do high impact presenters accomplish during an oral presentation?

   A. Develop a communication strategy

   B. Organize an argument

   C. Convey confidence through their delivery

   D. Integrate visuals

   E. Handle questions effectively

   F. All of the above

2. Which of the following is NOT one of the questions that should be answered prior to preparing slides for a talk?

   A. I want my audience to remember ____.

   B. I want my audience to write down ____.

   C. I want my audience to do ____.

   D. I want my audience to feel ____.

3. How many key points can audience members typically remember?

   A. 1–2

   B. 3–4

   C. 5–6

   D. 7–8

4. What is most important to an engineering argument?

   A. Deductive reasoning

   B. Justifying every decision

   C. Key features of the design

   D. Motivation for the design

5. When is the audience's attention the lowest?

   A. Thesis statement or key results

   B. During motivation

   C. After hearing "In conclusion"

   D. Details in the middle of the talk

6. Select the strong verbal transitions from the following. Select all that apply.

   A. First

   B. So

   C. However

   D. Next

   E. Therefore

7. Which of the following should you consider when selecting colors for a slide? Select all that apply.

   A. Choosing school colors to boost school pride

   B. Choosing culturally appropriate colors

   C. Choosing high contrast colors

   D. Choosing coherent themes

8. Sans serif fonts are best for _____, while serif fonts are best for _____.

   A. Writing; reading

   B. Projecting; printing

   C. Large audiences; small audiences

   D. Printing; projecting

Exercise #1

Introduction

A team must give a technical presentation regarding the design of a backpack for children with musculoskeletal disorders. The team has only completed the first three steps of the engineering process. Some members of the team have come up with the slide titles shown below (Figure C.7); however, the entire presentation has yet to be organized.

Need New Backpack Design	User-Defined Scale for Ease of Use	Existing Backpack Design	Backpack Design Criteria
Statement of Design Project	Kids Lack Independence at School	Range of Motion for Kids with Musculoskeletal Disorders	Future Plans

Figure C.7: Technical presentation slides.

Task

Organize the slides in a manner that will keep the audience engaged.

Exercise #2

Introduction

Critique the following slides in Figure C.8 on formatting, organization, and content. Be sure to note if any information may be missing and needed for clarity.

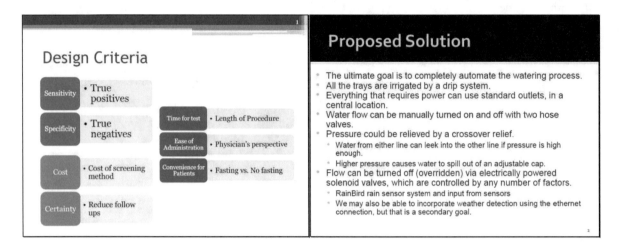

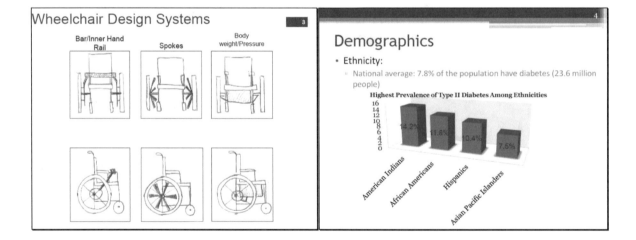

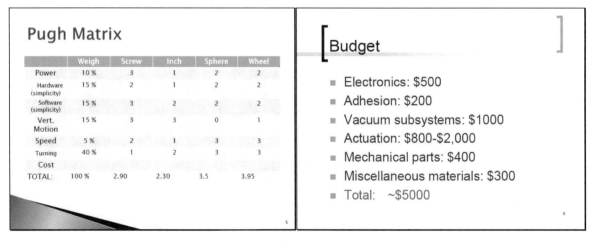

Figure C.8: Slides for critique.

Applying This Step to Your Project

Task: Outline Oral Presentation

Consider your current prototype and progress toward your design goals. Create an outline or script to present your brainstormed ideas, current solution, and/or prototypes. You should demonstrate the functionality of your prototype and future plans. Some topics to consider include the following.

- Motivation and clear statement of design problem,

- Some feasible design solutions,

- Process to evaluate design solutions,

- Final Pugh scoring matrix,

- Selected design solution,

- Prototypes,

- Lessons learned from prototypes, and

- Testing.

Each presentation should be 10–15 minutes long with an additional 2–3 minutes for questions from the audience.

SECTION C2: TECHNICAL MEMOS

Reading List

Read the following articles for an overview on the purpose and structure of technical memos. Learn the appropriate tone for memos and examples of content to include. Images, examples, exercises, and best practices were developed by Dr. Tracy Volz (Rice University), Dr. Linda Driskill (Rice University), and Dr. Liz Paley (Duke University).

Articles	https://www.nature.com/scitable/topicpage/memos-and-progress-reports-13987512/ (Online article)
	http://www.public.asu.edu/~thieme/Memos.ppt (Downloaded powerpoint)

Reading Notes

How are technical memos and e-mail messages similar? How are they different?

What information should be included in the introduction of a memo?

List at least four guidelines for formatting memos.

What are four recommended lines in a heading of a persuasive memo?

Content Summary

Technical memos are brief, written communications between individuals or teams, often in a technical professional setting. They can serve various purposes but are always designed to present information concisely and accessibly.

Most technical memos are written to communicate design decisions, such as establishing design criteria, selecting a direction for a prototype, or proposing tests. These design decisions are communicated using precise technical language, often with the support of numerical values. Decisions should be justified by clear technical reasoning. Before writing, you should determine the purpose(s) of your communication and what information to include to achieve that goal. Because memos are typically 1-3 pages, writers must communicate information clearly, concisely, and precisely. The following questions may help you plan your memos.

- To whom are you communicating?

- Why are you communicating (e.g., announcement, request, reply to inquiry)?

- What do your readers already know, and what do you want them to learn?

- What actions (if any) do you need from your readers?

- What tone will elicit the responses you desire?

Memo Organization Serves the Needs of Readers

Readers expect memos to flow in a logical order, beginning with the most important information and proceeding to least important. This "top-down" organization applies to the memo as well as to individual paragraphs (and often even to individual sentences). The opening paragraph of a memo summarizes the purpose and main points, while subsequent paragraphs provide supporting material. Likewise, topic sentences express the main ideas of the paragraphs while subsequent sentences provide further details.

Memo formatting is also designed to help readers easily locate, navigate, and process information. A detailed explanation of memo organization and guidelines for formatting is given in Table C.2. Helpful devices include:

- *Headings* that orient readers to the memo's purpose and provide at-a-glance essential context,

- *Section titles* that help readers quickly navigate the document,

- *Paragraph breaks* that help readers follow the logical flow of topics, and

- *Bulleted or numbered formats* that can make lists easier to read.

Visuals Support Claims

Well-designed figures and tables can sometimes summarize information more efficiently than prose. Use such visuals to support claims made in the text. Visuals do *not* stand on their own.

- Include numbers and captions *above* tables and *below* figures.

- Introduce visuals in the main text (e.g., "see Table 1").

- Explain in the text what readers should observe from the visuals (e.g., important data, results, conclusions).

Design and format visuals so readers can quickly cull essential information. Insert visuals immediately following the paragraph in which they are first mentioned, or in an appendix.

Writing Style Facilitates Flow

Readers appreciate prose that is clear, cohesive, and precise.

- Put key actions in verbs and keep verbs near their subjects.

  - Poor: "The cause of the cantilever's failure at supporting the load was [X]."

  - Better: "The cantilever failed to support the load because [X]."

- Begin sentences with old or familiar information, and push new, unfamiliar, or complex information to the ends of sentences.

  - Poor: "The collapse of a dead star into a point perhaps no larger than a marble creates a black hole."

  - Better: "Black holes are created by the collapse of a dead star into a point perhaps no larger than a marble."

- Use precise prose.

  - Poor: "The system must be light."

  - Better: "The cantilever must weigh less than 5 lb."

Citations Give Credit Where Due

If you quote, paraphrase, or otherwise rely on sources, cite them appropriately. Not only is this the right thing to do, it also lends authority to your text and enables readers to consult your sources for additional information if needed.

References

1. Driskill, L. (2000). Writing a technical memo (abridged). [Personal communication].

2. Grossenbacher, L., and Matta, C. (2014). Memos and progress reports. Retrieved September 5, 2017, from https://www.nature.com/scitable/topicpage/memos-and-progress-reports-13987512.

3. Williams, J. and Bizup, J. (2013). *Style: Lessons in Clarity and Grace*, 11th edition, Pearson, NY.

Table C.2: Technical memo style and formatting checklist	
Category	**Requirement**
Memo Organization	**Headings:** Includes properly formatted memo heading
	Paragraphs/section titles: Divides memo into appropriate paragraphs with section titles
	Topic sentences: Expresses main idea of each paragraph with topic sentence
	Flow: Sequences information logically; keeps content short and relevant
Clarity and Style	**Diction:** Uses clear, concise diction; eliminates empty qualifiers
	Consistent terminology: Keeps terminology consistent and precise throughout memo
	Definitions and acronyms: Spells out acronyms the first time they appear
	Active voice: Uses mostly active voice
	Mechanics: Uses correct spelling, grammar, and punctuation
Figures	**Captions:** Includes informative captions positioned below figure; numbers figures in order throughout document
	In-text references: References figures within text of document; inserts figures directly after paragraph in which they are first referenced or in Appendix
	Labels: Labels relevant features of images with consistent units and scale
	Quality: Uses high-quality images
	Legends: Includes legends, when necessary
Tables	**Captions:** Includes informative captions positioned above table; numbers tables in order throughout document
	In-text references: References tables within text of document; inserts tables directly after paragraph in which they are first referenced or in Appendix
	Headings: Uses consistent capitalization and formatting within table
Citations	**Reference list:** Uses proper citation format
	In-text citations: Uses consistent, ordered citations

Review Questions

1. When communicating within a company, which format of writing is best for presenting information concisely and accessibly?

    A. Technical memo

    B. Novel

    C. Research paper

    D. Essay

2. Technical memos use _____. Select all those that apply.

    A. Discipline-specific language

    B. Technical reasoning to justify claims

    C. Precise technical and scientific terms

    D. Clear and concise prose

3. Where should the most important information be placed in a technical memo?

   A. In the first paragraph

   B. In the last paragraph

   C. Spread throughout the memo

   D. In the subject line in the heading

4. What is wrong with this heading?

   To: Dr. David Ward
   From: Bubble C-PAP Team
   Date: Sept 23, 2018
   Re: Progress

   A. The date and subject lines should be aligned.

   B. The "to" line is too vague.

   C. The "re" line is too vague.

   D. The subject line should be in all caps.

5. What information should be included in the heading? Mark all that apply.

   A. Date

   B. Re (subject of memo)

   C. Name of organization

   D. To (to whom the memo is addressed)

6. Where should visuals go in a technical memo? Select all that apply.

   A. Immediately following the paragraph it is mentioned

   B. In the appendix

   C. Before the paragraph it is mentioned

   D. Right after being referred to (i.e. in the middle of the paragraph)

7. Which of the following tips should you use to structure your sentences in your technical memos? Select all that apply.

   A. Begin sentences with old/familiar information

   B. Keep key actions in verbs and keep verbs near their subjects

   C. Use simple vocabulary

   D. Use precise prose

Exercise #1

Introduction

Sample Technical Memo

U.S. Dept of Interior, Minerals Management Service, Gulf of Mexico OCS Region
Safety Alert No. 192
January 4, 2020
Contact: Jack Leezy 555-736-5555
Water Survival Craft

The Minerals Management Service (MMS) and the United States coast Guard (USCG) have examined a study conducted by an offshore operator of their survival craft payloads. This study looked at the manufacturer's design standards for several models of covered survival craft (lifeboats) used on the operator's offshore facilities. During this examination, the operator discovered that some of their survival craft were rated by the manufacturer to accommodate more personnel than the operator could properly seat within the unit. This was confirmed during several field trials. The operator concluded that this occurred because the "average" worker on their offshore facilities was larger than the approved design standard used to build the survival craft. The personnel weight standard used for the design of survival craft is 165 lb with a seat width of 14 inches. Although this standard may be appropriate for the offshore industry in other areas of the world, it may not be appropriate for personnel working in the Gulf of Mexico (GOM). The operator's study concluded that the average offshore worker at their facilities weighs 210 lb and has a seat width of 17 inches. Therefore, the difference in weight and width of the design standard and that of the personnel in the offshore industry in the GOM could have an effect on the seating capacity, stability, buoyancy, and structural adequacy of the survival craft. To accommodate the larger personnel at their GOM facilities properly, the operator chose to reconfigure some of the seating as well as de-rate the overall capacity of selected survival craft. The operator examined this same issue with respect to life rafts, but did not observe a similar problem.

Operators are advised to be aware of these potential overloading situations on survival craft caused by the larger size of the average GOM offshore worker. For more information, operators are urged to contact their nearest Coast Guard Officer in Charge of Marine Inspections.

Task

Revise the technical memo. What is the best order for the information below? List the order by number in the right column.

Operators should test carefully the effects of any reconfiguration on stability, buoyancy, and structural adequacy if changes are made. For more information, operators are urged to contact their nearest Coast Guard Officer in Charge of Marine Inspections or Jack Leezy at 555-736-5555.

_____

The Minerals Management Service (MMS) and the United States Coast Guard (USCG) examined a study conducted by an offshore operator, Williams Petroleum, of its survival crafts' payloads in the Gulf of Mexico (GOM). The manufacturer had rated some of the operator's survival craft to accommodate more personnel than the operator could fit in the unit. The "average" worker on its offshore facilities was larger than the approved design standard used to build the survival craft.

_____

TO: Operators of Offshore Platforms in the Gulf Coast
FROM: U.S. Department of the Interior Minerals Management Service
CONTACT: Jack Leezy at 555-736-5555
DATE: January 4, 2020
SUBJECT: Evaluate rescue craft for possible danger

_____

Table 1: Weight and seat width for average worker compared to design standard

_____

To accommodate the larger people at their GOM facilities property, Williams chose to reconfigure some of the seating, as well as lower the overall capacity rating of selected survival craft. Williams _____ examined this same issue with respect to life rafts but did not observe a similar problem.

	Weight	Seat Width
GOM average worker	210 lb	17 in
Design standard	165 lb	14 in

_____

These differences could affect the stability, buoyancy, and structural adequacy of the survival craft. Operators are urged to evaluate the load capacity of their water survival craft.

_____

Exercise #2

Introduction

Sample Technical Memo

Date: September 3, 20XX
To: Dr. Ann Saterbak
From: Team A
Subject: Lighting Structure for Engineers Without Borders Bridge in Matagalpa, Nicaragua
Problem Statement: Since the local chapter of Engineers Without Borders (EWB) constructed a bridge last year in the village of Matagalpa, Nicaragua, the bridge and surrounding area has become a hub for crimes such as theft, assault, and rape. Our goal is to design a public lighting system for the bridge in order to reduce these crimes. The criminal activity has made the bridge less accessible to local villagers due to the threat of criminals. A lighting system will effectively deter criminals.
The following features are needed within our design.

- Low cost: maximum cost of $25/light

- Durability: At least 5 years in local environment; resistant to vandalism and theft

- No recurring costs: Should not incur annual costs by using local utilities

In addition to the previous general features, the lighting system should include the following characteristics: an audible alarm that can be utilized if the villagers are about to be assaulted; an automated timing system or switch to save energy during the day; potentially replicable for surrounding villages; simple design for simple repairs considering the locals' engineering education; and well documented instructions for those building the prototype.

Team Structures and Activities
Team Name: Team A
Client: Project leader of a Nicaragua team in EWB responsible for building the bridge
Budget: $500
Deliverables:

- Written instructions by October 2nd

- Prototype and testing of the design by October 23rd

- Complete blueprints by November 29th

Task

Critique this technical memo. List several strengths and several weaknesses.

APPENDIX D

Project Planning

Task management is vital for completing an engineering design project. Planning a project helps ensure that progress is being made in a reasonable amount of time and that deadlines are being met. This appendix describes how to decompose a project into discrete tasks and create a timeline for accomplishing them.

Section D1: Work Breakdown Structure explains the process of creating a work breakdown structure. Tasks identified during a work breakdown structure act as a basis for a project timeline, allowing an engineering team to visualize all the smaller steps toward a final solution. Section D2: Planning Tools outlines several planning tools that can be used to organize tasks in a project. Regardless of which platform is used, it is important to map the project timeline, identify concrete tasks, and stay on task to complete the project.

SECTION D1: WORK BREAKDOWN STRUCTURE

Playlist

Watch this video playlist for an overview about work breakdown structures (WBS). Learn how to break down a large project into manageable subtasks.

Videos	bit.ly/additionalstepsd01-workbreakdownstructure

Video Notes

How are work breakdown structures visually displayed?

What are examples of products created from a WBS?

What are deliverables in a project?

When is a task sufficiently broken down?

Content Summary

A WBS is a visual tool that shows the hierarchical decomposition of tasks needed to complete a project. A WBS is organized around the major outcomes or products of a project (Figure D.1). To make a WBS, the team should decompose tasks by:

- Listing major tasks and

- Breaking major tasks into subtasks

Break tasks down until you can identify the lead person and estimate the length of time required to complete the subtasks. One possible WBS for the Safe Soap team is shown in Figure D.1.

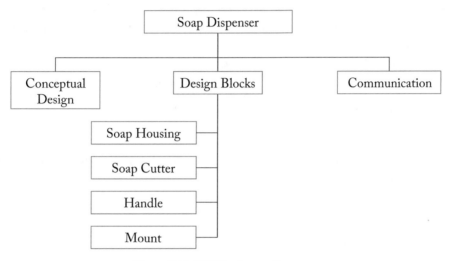

Figure D.1: **WBS** of soap dispenser.

Review Questions

1. Work breakdown structures are oriented around the ___ of a project.

   A. Deliverables

   B. Users

   C. Client

   D. Building material

2. Breaking tasks down to subtasks is helpful in determining ___.

   A. Who is responsible for what task

   B. What needs to be done for a project

   C. What processes are needed to carry out each task

   D. All of the above

3. Which of the following might dictate how far a team breaks down tasks and subtasks? Select all that apply.

   A. Estimates of the time required for each task

   B. The number of people required to complete a task

   C. The number of reports required

   D. Complexity of the problem

4. Two separate teams are working on the same project. Team A breaks the project into major tasks of design context review, prototype 1, prototype 2, and oral presentation. Team B breaks the project down into the major tasks of conceptual design, physical prototypes, and documentation. Which team's breakdown is better to use?

   A. Team A's breakdown is better because it includes more tasks.

   B. Team B's breakdown is better because it uses the term "physical prototypes" rather than numbering the prototypes. This gives them more flexibility.

   C. Neither breakdown should be used because neither includes the crucial first step of the EDP: clarify team assignment.

   D. Either breakdown is appropriate and could be used.

5. Work breakdown structures decompose a project in a _____ way.

   A. Sequential

   B. Linear

   C. Hierarchical

   D. None of the above

6. If teams cannot _____, they likely have not broken down a subtask far enough.

   A. Assign one member to a subtask

   B. Generate design ideas

   C. Write their first report

   D. Find online research resources

Exercise #1

Introduction

You are hosting a surprise birthday party for your friend in 2 weeks. At this party, you plan to cook a meal for everyone and play board games such as Monopoly and Clue. You will host the party in the common room of your dorm. You plan to invite twenty friends.

Task

Develop a work breakdown structure to plan this birthday party.

Applying This Step to Your Project

Task: Generate List of Tasks

Consider the major components of your design project and the specific steps needed to complete these components. A good starting point is to consider the tasks associated with each design block. You can also consider the tasks in the EDP, such as establish design criteria, brainstorm, or prototype. Be sure to consider practical tasks such as ordering materials and communicating with your client, if applicable.

Continue this discussion by making an exhaustive list of tasks for your project. You can use the hierarchy of a work breakdown structure (Table D.1), or brainstorm tasks in no particular order.

Task: Order List

After the exhaustive list has been made, re-order the list chronologically. Then, review the list to identify what is missing and add those tasks. Also, break up any tasks that are very large into smaller tasks so that you can estimate the time to complete the task.

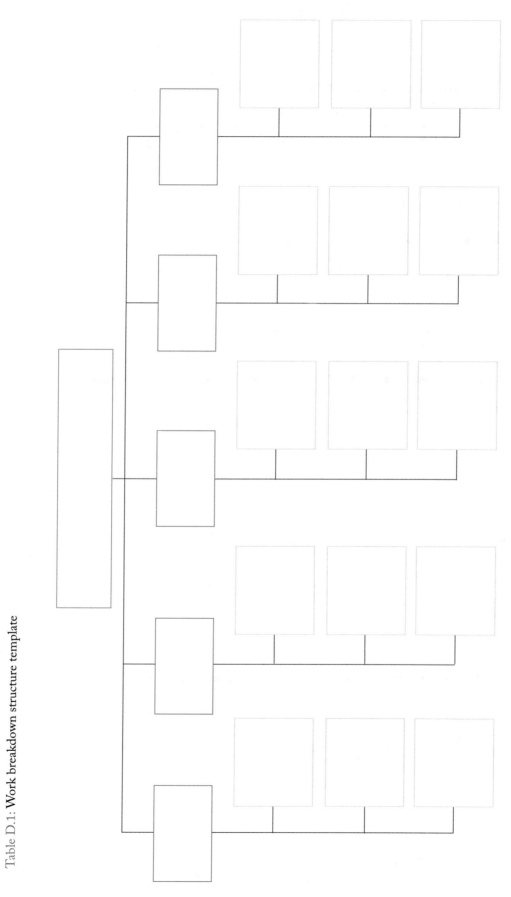

Table D.1: Work breakdown structure template

SECTION D2: PLANNING TOOLS

Playlist

Watch this video playlist for an overview about project scheduling tools. Learn how to create a Gantt chart and a Trello board.

Videos	bit.ly/additionalstepsd2-projectplanning
Videos	bit.ly/additionalstepsd2-trello

Video Notes

How are tasks ordered in a Gantt chart? In a Trello board?

How should the overall timescale be discretized for a 2-week project? For a 1-year project?

What is a milestone?

List two ways that color coding can be used for a Gantt chart or Trello board.

Content Summary

Both Gantt charts and Trello boards are project planning tools that organize tasks and subtasks generated from a WBS. These planning tools track tasks as a function of time and also note the team members who are assigned to complete the tasks. When creating a Gantt chart or Trello board, you need to keep the following in mind.

- Concurrency: Events or activities that occur at the same time (e.g., prototyping two independent design blocks can be done concurrently)

- Dependency: Event or activity that is dependent upon the completion of another event or activity (e.g., building a prototype with a new part is dependent upon ordering and receiving that new part)

- Lag: A delay or waiting period that results from an outside factor or situation

Gantt Chart

A Gantt chart is a tool used to describe the life cycle of a project (Figure D.2). In order to make a Gantt chart, your team must complete the following six steps.

1. Identify tasks and subtasks.

2. Define overall timescale and deadline.

3. Identify basic order of tasks.

4. Define relationships, like concurrency and dependency, between tasks.

5. Assign a lead member and any supporting members to tasks and subtasks.

6. Tie each activity with elapsed time period and due date.

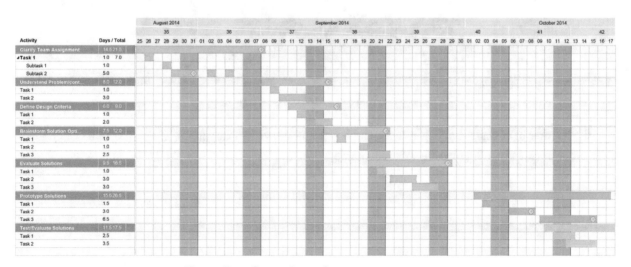

Figure D.2: Gantt chart of an engineering project.

Trello Board

A Trello board (Figure D.3) is a tool used to organize the tasks and subtasks created from the WBS. Team members and due dates are assigned to tasks on the Trello board. In order to make a Trello board, your team must complete the following steps.

1. Identify tasks and subtasks.

2. Create categories or lists that help your team organize different parts of the project. List headings could be deadlines, weekly goals, weekly progress, or other.

3. Create a card for each specific task or subtask. Place tasks and subtasks in the lists.

4. Define relationships, like concurrency and dependency, between tasks.

5. Assign a lead member and any supporting members to each card.

6. Tie each card with a due date.

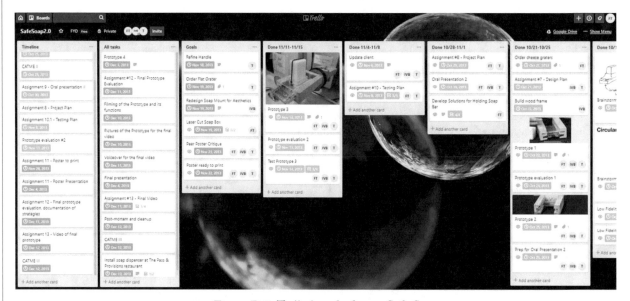

Figure D.3: Trello board of team Safe Soap.

The features of concurrency, dependency, and lag can be shown by categorizing cards in specific lists, rearranging the order of cards in a list, or using labels on cards.

Review Questions

1. What questions do Gantt charts and Trello boards answer? Select all that apply.

   A. What solution should I select?

   B. How long should a task take?

   C. What task is most important?

   D. What tasks need to be done?

2. On a Gantt chart, tasks are listed ____.

   A. In order of importance

   B. In chronological order

   C. In reverse chronological order

   D. According to team preference

3. The time required to complete a task is represented by the ____ of the bar.

   A. Length

   B. Fill color

   C. Width

   D. Texture/pattern

4. A dependency is a ____. Select all that apply.

   A. Waiting period that results from an outside factor or situation

   B. Task that is contingent upon another task

   C. Task that happens independently of another task

   D. Task that happens simultaneously with one another task

5. What does "Title A" refer to on this Gantt chart?

							Days								
Title A	Title B	1	2	3	4	5	6	7	8	9	10	11	12	13	14
Section 1	5														
Section 1	7														
Section 2															
Section 2															
Section 1	20														
Section 2															
Section 2															
Section 1	16														

A. Time block

B. Activity

C. Hours

D. Subtasks

6. In Trello, columns are typically used to keep track of ____, while cards are used to keep track of ____.

A. Time periods; tasks and subtasks

B. Dependencies and concurrencies; design blocks

C. Time periods; design blocks

D. Dependencies and concurrencies; tasks and subtasks

7. Every card should eventually have ____ assigned to it. Select two answers.

A. Team members

B. A label

C. A color

D. A due date

8. At the end of each week, what should your team do with your Trello board? Select all that apply.

A. Check your progress against your Trello board.

B. Review if the tasks set for the future make sense.

C. Update your Trello board with any new tasks that need to be completed.

D. Rearrange any cards in your Trello board as necessary.

Exercise #1

Introduction

Your team is developing a new coffee maker (Figure D.4). These systems work by turning on a heating mechanism that brings the water placed into it to a boil. Once the machine senses that the water has reached the correct temperature, it flows into a funnel-like structure containing a filter and ground coffee beans. The liquid that flows through this reservoir is collected by a decanter.

Up to this point, your team has created sketches of the design and has discussed the design in some detail. The team can clearly identify three discrete, decomposed aspects of the design: the water heater, the funnel, and the decanter. During the next month, your team needs to develop a series of successive prototypes. Table D.2 is the first iteration of your Gantt chart.

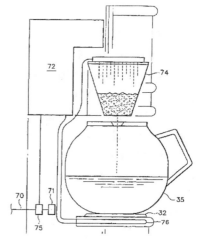

Figure D.4: Diagram of a coffee maker.

Table D.2: Project Gantt chart						
Activity	**Time**	**Week 1**	**Week 2**	**Week 3**	**Week 4**	
Complete CAD drawings	10 hr	▮				
Develop 1st prototype: heater	2 hr		▮			
Develop 2nd prototype: heater	2 hr			▮		
Develop 1st prototype: funnel	1 hr		▮			
Develop 2nd prototype: funnel	2 hr			▮		
Develop 1st prototype: decanter	1 hr		▮			
Develop 2nd prototype: decanter	2 hr			▮		
Integrate heater and funnel	½ hr				▮	
Integrate heater and funnel	½ hr				▮	
Test integrated system	4 hr				▮	
Revise CAD drawings	5 hr					▮
Develop 3rd prototype-integrated	5 hr					▮
Test 3rd prototype-integrated	10 hr					▮

Task

Describe two high-level strengths and two high-level weaknesses of the Gantt chart.

Specifically critique the Gantt chart. As appropriate, edit the Gantt chart to reorder tasks or include missing tasks; increase or decrease time for specific tasks; and address concerns around concurrencies, lags, and dependencies. You may write directly on the chart.

Exercise #2

Introduction

Your team is developing a new coffee maker. These systems work by turning on a heating mechanism that brings bring the water placed into it to a boil. Once the machine senses that the water has reached the correct temperature, it flows into a funnel-like structure containing a filter and ground coffee beans. The liquid that flows through this reservoir is collected by a decanter.

Up to this point, your team has created sketches of the design and has discussed the design in some detail. The team can clearly identify three discrete, decomposed aspects of the design: the water heater, the funnel, and the decanter. During the next month, your team needs to develop a series of successive prototypes. Table D.3 is the first iteration of your Trello board.

Table D.3: Project Trello board

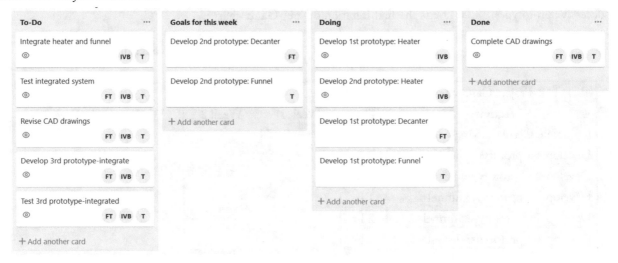

Task

Describe two high-level strengths and two high-level weaknesses of the Trello board.

Specifically critique the Trello board. As appropriate, edit the Trello board to reorder tasks or include missing tasks; develop possible due dates; address concerns around concurrencies, lag, and dependencies; alter general organizational structure; etc. In other words, edit the Trello board to improve it.

Applying This Step to Your Project

Task: Generate Gantt Chart or Trello Board

Using your work breakdown structure, create a Gantt chart or Trello board. The chart or board should include all tasks and subtasks (e.g., prototype development, testing, and documentation), along with their respective deadlines. This will make it clear which tasks need to be prioritized and completed by certain dates. Concurrent and dependent tasks should be noted. Both Gantt charts and Trello boards should identify the leader and supporting personnel for each particular task.

Be as specific as possible with the scheduling of steps. As an example, pretend your team identified its design as being made of three different design blocks. Your Gantt chart or Trello board should then reflect the sourcing, purchasing/acquiring, prototyping, assembly, testing, iteration, and documentation of each of those three design blocks. Keep in mind that the planning tool you use should actually be a guide to manage and prioritize your work, as well as stay on time.

Note that your Gantt chart or Trello board may need to be updated to reflect project progress.

Authors' Biographies

Ann Saterbak is Professor of the Practice in Biomedical Engineering and Director of the First-Year Engineering Program at Duke University. Since joining Duke in June 2017, she launched the new *Engineering Design and Communication* course. In this course, first-year students work in teams to solve community-based, client-driven problems and build physical prototypes. Prior to Duke, she taught at Rice University, where she was on the faculty since 1999. Saterbak is the lead author of the textbook *Bioengineering Fundamentals*. At Rice and Duke, Saterbak's outstanding teaching has been recognized through five school- and university-wide teaching awards. For her contribution to education within biomedical engineering, she was elected Fellow in the Biomedical Engineering Society and the American Society of Engineering Education. She is the founding Editor-in-Chief of *Biomedical Engineering Education*.

Matthew Wettergreen is the Director of the Global Medical Innovation Master of Bioengineering program at Rice University. He is also an Associate Teaching Professor at the Oshman Engineering Design Kitchen at Rice where he teaches engineering design and prototyping. Based on a deep interest in curriculum that builds capacity for student development in makerspaces, he has co-created materials and delivered workshops to establish international engineering design programs. Wettergreen is the faculty mentor for Rice's *Design for America* chapter, for which he has been awarded the Hudspeth Award for excellence in student club mentoring. For his contributions to the development of the design curriculum at Rice, he received the Teaching Award for Excellence in Inquiry-Based Learning. His design work has been featured in *NASA Tech Briefs*, the *Wall Street Journal*, *Make Magazine*, *Atlantic Monthly*, and *Texas Monthly*.

Printed in the United States
by Baker & Taylor Publisher Services